A CENTURY OF FISHING

Fishing from Great Yarmouth and Lowestoft 1899-1999.

Compiled by MALCOLM WHITE

Proceeds from the sale of this book will be shared between the LFSBS, RNMDSF and the Lowestoft & East Suffolk Maritime Heritage Museum.

Front Cover Photograph
The drifter/trawler *YH73 Rose Hilda* was built in 1930 by Fellows & Co. Ltd., at Great Yarmouth for W. J. E. Green Ltd. She is seen here leaving the harbour at Great Yarmouth. In 1954, *Rose Hilda* was sold to F.E. & M.O. Catchpole of Lowestoft and became *LT90 Dawn Waters*. Her F.W. Carver 33hp triple expansion steam engine and associated boiler, were replaced by a 6cyl. 440bhp Crossley diesel engine. The conversion was undertaken at Lowestoft by LBS Engineering Co. Ltd. During 1960 the ownership of *Dawn Waters* was transferred to Gilbert & Co. Ltd. She left Lowestoft in 1962 for Milford Haven, after being sold to Norrard Trawlers Ltd. *Dawn Waters* was sold for scrap in 1970.

Title Page Photograph
The latest addition to the Lowestoft trawling fleet, *LT1005 St. Anthony*, arrived at her home port during the late afternoon of the 27th. March 1999. The order for her building was placed by Colne Shipping Co. Ltd. with Maaskant Shipyards of Stellendam in the Netherlands. It was confirmed on the 28th March 1998. At the time of her commissioning, she was described as the United Kingdom's most advanced "state of the art" supertrawler. *St.Anthony* is 42 metres in length and powered by a 2000hp Deutz engine. Much of her construction was carried out in Poland, and she was the third of a series of three completed at the Dutch shipyard. The other two new vessels were built for a Dutch fishing company. The *St. Anthony* replaced the *St. Vincent* and *St. Simon*, which were sold, in the Colne fleet.

Back Cover Photograph - Left
Built by Richard Dunston (Hessle) Ltd. in 1965, *LT508 Boston Wayfarer* is a good example of the smaller vessels owned by Boston Deep Sea Fisheries Ltd., and their subsidiary companies. She was powered by a 528hp 8cyl. Blackstone diesel engine and was 94ft. in length. *Boston Wayfarer* was transferred a number of times between Boston subsidiaries, namely St. Andrews Steam Fishing Co. in 1966, B.D.S.F. in 1968, W.H.Kerr (Ship Chandlers) in 1974 and Onward Fishing Co. in 1979. In 1981 she left Lowestoft for Capetown and new ownership. She was registered at Luderitz as L404.
This fine photograph of *Boston Wayfarer* was taken on 2nd. February 1974.

Back Cover Photograph - Right
The stern trawler *LT144 St.Phillip* was built in 1975 by Richards(Shipbuilders) Ltd. at their Great Yarmouth shipyard. This excellent view of the stern of *St.Phillip* was taken on the 28th. June 1976 as she entered Lowestoft, on return from a fishing trip. *St.Phillip* has had a considerable number of changes of ownership. During 1986 she became *Kerry Kathleen*. In 1989/90 she was converted to a safety standby vessel, and had her original name reinstated. The *St.Phillip* was sold again in 1998 and became *Viking Vulcan*.

Charities which will benefit from the sale of this book are as follows :-

Lowestoft Fisherman's and Seafarers Benevolent Society (LFSBS)

Royal National Mission To Deep Sea Fishermen (RNMDSF)

Lowestoft & East Suffolk Maritime Heritage Museum

Published by	Malcolm R. White,
	71 Beeching Drive,
	Denes Park,
	Lowestoft,
	Suffolk,
	NR32 4TB
	England,
	United Kingdom.
First Published	1999
Reprinted	2000
Copyright	© Malcolm R. White 1999
ISBN	0 9532485 1 8

All rights reserved.

No part of this publication may be reproduced, stored in a retrieval system, or transmitted in any form or by any means electronic, mechanical, photocopying, recording or otherwise, without the express permission of the Publisher and the Copyright holders in writing.

Please note that reproduction fees are demanded on certain photographs by the copyright owner.

Printed by	Micropress Printers Ltd.,
	27 Norwich Road,
	Halesworth,
	Suffolk,
	IP19 8BX,
	England,
	United Kingdom.

In Scottish waters, the 1918 built "standard" steam drifter *LT237 Jacketa.*

CONTENTS

	Page		Page
ACKNOWLEDGEMENTS	4	THE PHOTOGRAPHIC REVIEW	
INTRODUCTION	6	a) 1899 - 1990s	12
EAST ANGLIAN FISHING	8	b) AT THE END OF THE TWENTIETH CENTURY	182
		BIBLIOGRAPHY	184

ACKNOWLEDGEMENTS

The following organisations, companies and representatives of companies have provided greatly appreciated assistance and co-operation. In some cases I have been allowed to use material for which they own the copyright.

Malcolm Berridge	George Catchpole	Peter Catchpole
Robert Catchpole	John Hashim	Antony Jarrold
Tim Oliver	Brian Ollington	Richard Reeve
Hugh Sims	John Wells	Trevor Westgate
David White	Elizabeth White	Fishing News
Putford Enterprises Ltd.	Colne Shipping Co. Ltd.	Tate and Lyle plc

Lowestoft Fishing Vessels Owners Association

I am most grateful to John Wells and his staff for help and assistance with archive material used in this publication, and also for access and use of photographs from the John Wells Heritage Collection.

My personal thanks to the many kind people from Lowestoft, Gorleston and Great Yarmouth who have helped with research, and provided information in connection with this book. Much appreciated is the support and assistance from members of the Port of Lowestoft Research Society, and also from Peter Parker, the Chairman of the Lowestoft and East Suffolk Maritime Society,

For undertaking the task of checking the text I am very much indebted to Stuart Jones BA., formerly of the CEFAS Laboratory, Lowestoft.

PHOTOGRAPHIC OWNERSHIP AND COPYRIGHT

The vast number of photographs used in this publication come from many sources, some were taken by the author and many are from his collection.
With much appreciation, a large amount of copyright material has been included from the archives of Richards (Shipbuilders) Ltd. (Cyril Richards Collection), the archives of Brooke Marine Ltd. (Peter Hansford Collection, Brooke Archivist), and the Port of Lowestoft Research Society. Also with their kind permission, items for which copyright is held by the following have been included in this book:-

Jarrold & Sons Ltd., Scottish Fisheries Museum, Shetland Museum, Lowestoft & East Suffolk Maritime Society, Eastern Counties Newspapers Ltd., Waveney District Council, Maritime Photo Library, John Allsop, Ken Carsey, Stanley Earl, Pamela Graystone, Ernest Harvey, Kenneth Hemp, Peter Jenkins, Ken Kent, Peter Killby, Brian Ollington, Eileen Pembroke, Jack Rose, Ivonne Revell, Neil Watson, Parry Watson, Pauline White.

The following bodies have provided support and assistance in the preparation of this book:-

Scottish Fisheries Museum	Shetland Museum
Shetland Islands Council	National Fisheries Heritage Centre
Port of Lowestoft Research Society	Lowestoft War Memorial Museum
Great Yarmouth Maritime Museum	Suffolk Record Office

Lowestoft & East Suffolk Maritime Heritage Museum

ABOUT THE AUTHOR

Malcolm White was born and has lived all his life in Lowestoft. He comes from a family with long associations with the sea, and the fishing industry. His grandfather, was a Lowestoft smacksman and his mother was a beatster. He has fond memories of sitting by the coal fire at home, on herring nets which his mother would mend for the fishing companies of Beamish, Hobsons and Offord.
For the majority of his life, Malcolm has been associated with the maritime aspects of Lowestoft. He spent many years working as an electrical engineer for a major local fishing company, and has also worked in local shipyards, marine engineering works and the Lowestoft Harbour railway works. Whilst maintaining the undersea cables which terminated at Lowestoft, Winterton and Covehithe, Malcolm gained substantial experience in dealing with cableships of various nationalities and the damage caused by anchors and beam trawls to these cables. Now retired he is currently undertaking major research projects on East Anglian maritime heritage. Malcolm is a member of many maritime, historical and transport societies and gives slide and film shows of local maritime and transport interest. A member of the publishing committee of the Port of Lowestoft Research Society, he has written a number of articles on maritime matters for magazines and technical journals.
The first major book focusing on the harbour, fishing and shipbuilding aspects of Lowestoft, "Down The Harbour 1955 - 1995", was written and published in 1998 by Malcolm.

Postcards depicting the fishing industry at the beginning of the twentieth century. The cards on the left are from Gt. Yarmouth and those on the right from Lowestoft.

INTRODUCTION

When their lunch, dinner or supper consists of fish, many people probably never stop to think of what is involved in catching fish, or what a vast organisation is necessary to make sure there is always a supply of fresh fish in the shops. A very important part of that organisation is the fishing industry. This has a very long history and was at one time one of the largest in Britain. In addition to the fishermen, large numbers have been employed at the ports in unloading, filleting, processing and ancillary trades such as ice making and net making and repairing. At the end of this supply and distribution chain has been the rail and road transport links, wholesale markets and shops.

Thousands of fishermen have put out to sea in fishing vessels of all kinds, men ready to face extremes of hardship and danger in order to maintain a supply of fish; risking their lives in often appalling weather conditions. In the late nineteenth century and into the twentieth century many of these were sailing vessels, often quite small and totally at the mercy of the elements, at sea in all conditions, in most cases with only the wind and tide as power sources. In the early part of the period considered in this book, these vessels would have none of the safety, communication and navigational aids taken for granted today. Over the years the design of the vessels working the North Sea from Gt. Yarmouth and Lowestoft has changed in a quite spectacular manner. The changeover from sail to steam meant that vessels could travel much further more quickly with consistent and reliable power. The further change in propulsion to motor power meant that vessels could be operated more cheaply with greater flexibility and in most cases, a major increase in the power available for fishing. This also gave a greater catching capability and a further increase in speed not available with previous power sources. With all today's technological aids, and more robust vessels with greater engine power, the sea and what it is capable of, remains as unpredictable as ever. The danger and hazards of the lifestyle of those who go to sea, are still there, but are now made a little more comfortable by the features and build of the latest vessels.

For a substantial part of the last hundred years a major contribution to the British fishing industry has come from the east coast ports of Lowestoft and Great Yarmouth working mainly, but not totally, the North Sea. From these ports, the principal types of fishing employed in 1899 and for the majority of the next one hundred years, were drift netting and trawling. Longlining has also taken place.

Drift net fishing was used to catch pelagic fish which tended to feed near the surface in shoals. Herring and mackerel were the most common types in the North Sea. This method of fishing is no longer used, except by longshore boats. Trawling is the method of catching what are known as demersal fish, common types of these being cod, halibut, plaice and sole.

In this book some of the fishing vessels seen, during the one hundred year period between 1899 and 1999, working from these two great fishing ports are reviewed. Some aspects of the supporting trades and occupations are also included. With regard to sailing fishing vessels, a " smack " in this book refers to a first class sailing trawling vessel. This is traditionally the local meaning of the name. A first class vessel was defined by the Customs Commissioners Act of 1786 as a decked boat or ship of more than 15 tons burthen. Many sources refer incorrectly to all sailing fishing vessels, including drifters, as smacks. The English sailing drifter was known as a "lugger", even though in later years they were not lug rigged but had adopted a gaff ketch rig.

Throughout this book the greatest emphasis has been placed on reviewing the period which will be within the living memory of the majority of the readers. An assumption has been made that the reader has an appreciation of trawling, drifting and longlining, however, a very brief appraisal of these is included for those not familiar with fishing methods.

During the one hundred years reviewed by this book, the everyday life of the two ports would feature a varied and random selection of vessels. This aspect has been applied to the contents of this book.

It is intended to let the photographs tell the story rather than have long chapters of text.

Malcolm White
Lowestoft
June 1999.

The harbour entrance at Lowestoft with vessels departing for the fishing grounds. The two nearest vessels are *LT226 Mary Louisa* on the left, and *LT349 Young Admiral* on the right.

EAST ANGLIAN FISHING

By the end of the nineteenth century it could be said that the ways in which the fishing industries at Great Yarmouth and Lowestoft had evolved, were in many respects similar. Large fleets of sailing trawlers and sailing herring drifters could be found at both ports with all the necessary support facilities. A number of converter vessels, able to become a trawler or drifter, and many inshore fishing boats were also to be seen. Earlier in the nineteenth century a number of fisherman, and later fishing companies, had moved to Great Yarmouth and Lowestoft from such ports as Ramsgate, Barking and Brixham, bringing their fishing vessels with them. This resettlement was due to a number of factors, one of which was to be nearer the rich North Sea fishing grounds. These newcomers brought with them new ideas which were to influence many aspects of the local fishing industry in future years.

The building of fishing vessels was carried out at various yards in both towns, and full repair facilities existed to service the large fleets. In both ports the herring fishing industry and the trawling industry were very well established with Great Yarmouth, the greatest herring port in the world, and one of the largest trawling centres in this country.

A century later, in 1999 the picture is very different. The trawling industry no longer exists in Great Yarmouth, having gone into rapid decline around the turn of the century. Despite attempts to revitalise it, including those made since the Second World War, establishment of the port again as a major trawling centre, has not materialised.

Herring fishing together with the vast industry and the associated employment it supported has totally disappeared from both ports. As a reminder of the town's great fishing past, Great Yarmouth has a substantial inshore fishing fleet.

At Lowestoft the trawling industry has survived, having gone through a period of vast restructuring brought about by many factors, some of which originate from the European Union and the Common Fisheries Policy. A number of large beam trawlers operate from the port, which also has many inshore fishing vessels.

Section 1 - DRIFTING AND DRIFTERS

Drift net fishing was used to catch pelagic fish; common types found in the North Sea have been mackerel and herring. Normally massing on the bottom during daytime, they would rise to the surface as the daylight faded at the end of the day, to feed. It was during the time that the fish were rising up to the surface, or in the vicinity of it, that the mature fish could get caught in the nets of the drifters. This method of fishing had changed little since being introduced by the Dutch before the sixteenth century.

The basic principal of drift net fishing is that once the nets are shot by the drifter into the sea, they would form a perpendicular barrier or wall in the water. It was hoped that the fish would swim into this wall and get caught. The herring drifters of the twentieth century could carry if required a fleet of between sixty and ninety nets, depending on the size of the drifter. Each net was nominally thirty three yards long and twelve yards in depth and could create a wall about two to three miles long and twelve yards in depth. The wall effect was achieved by the use of buffs (buoys made of canvas or plastic) which floated on the surface, and corks at the top or upper edges of the net. The upper part of the nets were joined to the buffs by ropes called buff strops, and there was one buff at each end of each net. Ropes called seizings were used at the bottom of the net to fasten it to a warp which passed along under the bottom of the nets. This warp was known by a number of different names varying from locality to locality. It was attached to a heavy rope in the vicinity of the drifter called the tissot, this was attached to the drifter and took the strain of the weight of the nets and their contents. In some localities the strop and the tissot were referred to as the buoy rope and the guy rope respectively.

Traditionally shoals of herring had moved down the east coast, and in the autumn were to be found in the southern North Sea. Every year for well over a hundred years until the late 1950s, large numbers of herring drifters, mainly Scottish, arrived at the East Anglian ports at this time. These had followed the shoals down the North Sea landing the fish caught on the way at such ports as Whitby, Scarborough, Grimsby, North Shields and Hartlepool. Many local drifters had several months earlier, travelled north to fish for herring from Scottish and North of England ports. The visiting drifters complemented the local drifters with the result that a huge fleet of herring drifters was to be found working for a few months at the end of each year from the ports of Great Yarmouth and Lowestoft. The larger proportion was to be found working from Great Yarmouth.

Towards the end of the nineteenth century the majority of the fleet consisted of sailing drifters. The Scottish drifters, depending upon their construction, were known as zulus or fifies. Their English counterparts were known as luggers. To maintain production away from their home base the Scottish merchants and curers followed their fleet down the North Sea. They used very labour intensive methods of production and employed many women to clean, split and pack the herring into barrels within layers of salt. These "fisher girls" arrived from Scotland at both towns by train. Called " girls " irrespective of their age, they worked at other English ports such as Scarborough when herring was being landed by the Scottish fleet.

In addition many other support workers arrived to work in the fish houses, gutting yards, net stores and smokehouses. This was an extremely busy time for such tradesman as barrel makers(coopers), beatsters, basketmakers and ship chandlers. The demands upon the services of the coal, oil, ice and salt merchants was continuous and the missionary and welfare services were also in demand. Very essential at this hectic time was the service provided by shipping grocers and butchers. Local businesses such as W. B. Cooper Ltd., John Devereux & Sons, Trawler and Drifter Supply Store, K. E. Hutsons, Ronnie Cook and Sydney Cook & Son were involved in this major provisioning exercise. The visits to Great Yarmouth and Lowestoft of the

Scottish "fisher girls" and the other various Scottish supporting trades lasted well into the 1950s.

The large scale replacement of the sailing drifter by steam powered vessels commenced just before the end of the nineteenth century, with the 1900s being marked by the rapid rise in popularity of these new vessels. The first local purpose built steam drifter was *LT718 Consolation*, launched in March 1897 for G. Catchpole of Kessingland near Lowestoft. It was built at the Chambers and Colby yard at Oulton Broad. The advantages of steam over sail soon became apparent and within a very short time large numbers of these steam drifters were being built for both English and Scottish owners at many different yards. A few sailing drifters however, remained in service until the early 1910s. Although not in great numbers the very first motor driven fishing vessels appeared about this time with the drifter *LT368 Pioneer*, built in 1901 at Reynolds yard at Oulton Broad, which was one of the more successful vessels. It was fitted with a four cylinder four cycle Globe Marine Gasoline engine built by the Pennsylvania Iron Works Co. of Philadelphia, U.S.A. It left Lowestoft in 1913 after being sold, bound for Spain. Some of the early motor powered fishing vessels such as *LT1035 Thankful* had a chequered history. By 1908 the *Thankful*, had been converted from motor to steam power.

North of the border, *ML30 Pioneer* was built for the Scottish Fisheries Board in 1905 at Anstruther, for demonstration purposes. Powered by a 25hp single cylinder four stroke Dan motor she toured various ports to show the new propulsion method to fishermen and owners. She retained her full set of sails but possessed many novel features. June 1905 saw her venture down to London, where she was inspected by M.P.'s and many interested parties. The same year she took part in the East Anglia herring fishing. Many consider the first really successful Scottish motorised fishing vessel to be the Eyemouth fifie *BK146 Maggie Jane's*, built in 1901 and fitted in 1907 with a 55hp Gardner 3KM engine. This vessel was so successful that many owners decided that the future was with motor and in 1908 four vessels all registered at Berwick were also fitted with Gardner engines. Thus started a trend in many Scottish ports to motorise several zulus and fifies. The first full powered Scottish motor drifter was *FR61 Gardner*, built in 1909 at Sandhaven. In East Anglia many sailing fishing vessels were also motorised and in 1928 Richards Ironworks at Lowestoft built the diesel powered drifter *Veracity*, fitted with a German engine. At that time, the steam drifter was very well established along the east coast ports of Scotland and England. It was to be many years before any other form of propulsion would provide a serious challenge. In 1961 however, the last operating steam drifter, *YH105 Wydale* was sold for breaking up in Holland. The remaining drifters were either relatively new diesel powered vessels, former steam powered drifters converted to take diesel engines or former Admiralty Motor Fishing Vessels (MFVs). The majority of the new build vessels were constructed at the Lowestoft shipyard of Richards Ironworks Ltd., between 1949 and 1960 for owners in both towns. These vessels were capable of operating as a drifter or trawler. When diesel finally ousted steam, the East Anglia herring season was but a tiny shadow of its former self and by the mid 1960s was no more.

The East Anglian herring fishery was at one time the greatest fishery in the world, generally accepted as peaking in 1913. In that year, 1163 Scottish vessels were at Great Yarmouth and Lowestoft, with the local drifters this made a combined fleet of 1683 vessels.

Monday 12th October 1913 proved to be a day of very heavy landings with four drifters landing over three hundred crans each and thirty drifters each landing over two hundred crans. Many others landed catches in excess of one hundred crans. A cran was officially considered as being 1320 herrings. Many people however, refer to a cran as 1000 herrings. The total landing for that season is recorded as 1,484,249 crans worth £1,404,232. Eighty-seven per cent of the catch that season was either cured or pickled and then sent to Germany and Russia.

The herring fishery is now but a memory, along with the hundreds of sail, steam, motor or diesel powered herring drifters which took part in the fishing from both ports. At Great Yarmouth the decline in the amount of herring caught and a diminishing market meant that by 1963, herring fishing from the port had effectively ended. At Lowestoft it lasted a little longer. The last Great Yarmouth drifters moved to Lowestoft and under new ownership, worked a little longer herring fishing with Lowestoft registrations. Eventually pair trawling for herring was tried out by both English and Scottish vessels from Lowestoft. In their last years, before being sold away from East Anglia the remaining drifters, undertook trawling for white fish as drift net fishing and pair trawling for herring finally became totally uneconomic. A number of steam, motor and diesel drifters did use other fishing methods in addition to drifting and trawling. Longlining and seine netting were two types of fishing which the drifters would undertake when economic circumstances dictated.

The last principal companies to own drifters in East Anglia were Bloomfields Ltd. of Great Yarmouth and Small & Co.(Lowestoft) Ltd. of Lowestoft. Other local drifter owners in the late 1950s and 1960s include the Eastick family, Gilbert & Co., Silver Fishing Co., G.M.Haylett, F.E.Catchpole, J.J.Colby, W. & L.Balls and A. W. Easto. The final seven former Great Yarmouth vessels and the remaining original Lowestoft drifters were later sold, finding new homes and a new lease of life in such places as Greece, South Africa, Italy, Scotland and Holland. Some were used for fishing. Many eventually became safety standby vessels for the offshore oil and gas industry, with others doing survey and support work, or acting as diving tenders. The very last herring drifter to operate from the East Anglian ports was *LT382 Wisemans*. Formerly the Banff drifter *BF154*, this vessel finished fishing from Lowestoft in November 1968.

The last English drifter/trawler built was the 147 ton *LT671 Suffolk Warrior*, completed by Richards Ironworks at Lowestoft in 1960. She was lost on the 15th June 1969 after a collision in the North Sea.

In 1999, very few former East Anglian herring drifters still exist, most have been lost at sea or broken up. This includes the majority of the last generation of diesel powered drifter/trawlers, built locally between 1949 and 1960. Amongst the survivors in this country at the time of writing, is the 1959 built former drifter/trawler *LT277 Valiant Star*.

Section 2 - TRAWLING AND TRAWLERS

The trawl is the most important and effective instrument for the capture of the "demersal" species of fish i.e. all those which unlike the herring or mackerel, live on, or very near, the bottom of the sea. It is also the most expensive method used for catching fish. Various types of gears and rigs are worked for small and large trawls, but irrespective of size they are all of the same principal. The trawl resembles a conical shaped bag of net with a wide mouth. Methods of keeping the mouth of the net open vary according to the type of trawl used. The beam and otter trawls have been the types principally used for the last one hundred years.

The beam trawl differs from the otter trawl in that the mouth of the net is kept open by a rigid beam of wood usually made of ash, elm, beech or oak. Modern trawlers use twin beams of steel construction. The otter trawl was provided initially with structures made of wood and metal, later to become totally metal, called otter or trawl boards. These kite-like boards are attached to each side of the trawl and equipped with iron brackets which, being specifically arranged so that each board diverts in an outward direction when towed, produce and maintain to the full extent, the mouth of the trawl. In order to increase the effectiveness of the otter trawl it was later modified to increase its catching ability and generically became known as the Vigneron - Dahl type. This type of trawl proved very successful when used with steam and diesel trawlers. These possessed the necessary consistent power to tow the trawl effectively along the sea bed. The beam trawl, being equipped with iron angular fittings at each end of the beam maintains the full spread, and should the sailing vessel become becalmed and towing speed reduced, the trawl would still remain in a fishing position. Under these circumstances the trawl boards of an otter trawl may settle horizontally on the sea bed with collapse of the net structure. As the sailing trawlers with their beam trawls were taken out of service and steam trawlers took over, the otter trawl became the principal way of trawling. It was used for the majority of the period covered by this book. In the 1980s, the preference of otter trawl over beam trawl was reversed for the Lowestoft trawlers operated by the three remaining owners. For over sixty years the trawlers using otter type trawls had been the mainstay of the fleet. These vessels were totally disposed of, and replaced for economic reasons with trawlers built and equipped for beam trawling. The beam trawl has once again become the principal form of trawling for the large trawlers of Lowestoft.

A vast sailing trawler concentration once existed at Great Yarmouth. Amongst them was the Short Blue Fleet with over 200 vessels. The Short Blues were owned by Hewett & Company of Barking, and took their name from the plain square blue flag flown by every Hewett ship.

The history of the Hewett company is very interesting. The company could trace its origins back to 1763 and were chiefly responsible for the growth of Barking as a major fishing port. Later they were to move their smacks to Great Yarmouth. After the First World War they based steam trawlers at Lowestoft and Fleetwood complete with London fishing registrations. In the 1930s, they combined with two other trawler fleets from companies which had ceased operations. Heward Trawlers Ltd. and a number of subsidiary companies were formed. As a result of this, in the late 1930s, the famous Gamecock and Red Cross Fleets of steam trawlers combined with the Short Blue fleet. The last two steam trawlers based in East Anglia and belonging to Hewett subsidiary companies were *LO200 Junco* and *LO251 Warbler*.

Much of the fish caught by sailing trawlers from the Great Yarmouth fleet was transferred at sea to fast steamships which took the fish direct to markets such as Shadwell and Billingsgate in London. These fleets stayed at sea for up to 10 weeks and comprising of many vessels, operated as one unit on the fishing grounds under the control of an admiral on a flagship. Control of the complete fishing operation was under the direction of the admiral using signal flags. Sailing trawlers from the Lowestoft fleet did not generally follow this practice and worked independently, tending to land their catches at their home port, the fish then being despatched by railway to many inland towns and cities. For many years, until the advent of the lorry, the railway was to be the main form of distribution for all fish landed, including the seasonal herring, caught during the Autumn. Locally the sailing trawler, depending on her size, could be referred to by a number of names some of which are smack, tosher or wolder. The nineteenth century saw a number of sailing fishing vessels classed as converter smacks, these could take on the role of sailing drifter or trawler as required by the owner when circumstances were most favourable to him(or her). Around the turn of the century the side fishing steam trawlers, working with otter trawls appeared. These worked side by side with the sailing trawlers using their beam trawls for many years. During the 1920s and 1930s large numbers of the sailing trawlers were broken up or sold, many going to Norway and Sweden. However, at the end of the 1920s over 120 of these sturdy, seaworthy magnificent vessels are recorded as still working from Lowestoft. Up until the outbreak of the Second World War in 1939, the following eight sailing trawlers were working from Lowestoft :- *LT744 Wave Crest, LT 823 Lustre, LT1099 Northern Queen, LT1130 Girl Edna, LT1203 Master Hand, LT1241 Boy Eric, LT1270 Our Need* and *LO401 Sir William Archibald*.

Trawling at Great Yarmouth, although once a major industry with many hundreds of sailing trawlers was not as successful in the twentieth century as it was at Lowestoft. However, for several years after the Second World War trawl landings were made at Great Yarmouth, the first in March 1946 by the drifter/trawler *LT344 Lord Anson* with Skipper J.Oakes in charge. Several locally owned large steam trawlers including *H541 Avon, GY397 St.CLair, A179 Scot, GY411 Nacre* and *GY383 Ugie Bank* worked out of Great Yarmouth during this period. In the early 1950s, large Jewish fishing vessels could be seen in the port. Several Polish fishing vessels operated from Great Yarmouth between 1947 and 1950. This fleet consisted of former minesweepers such as *MMS1045*, *MMS1047* and *MMS1028*, and former Lowestoft drifter/trawlers such as *Jenny Irvin*, *Mace*, *Acorn*, *Lord Hood*, *Lord Barham* and *Lord Collingwood*. These former Lowestoft vessels were renamed in the "*Dal*" series, *Dal 1*(*Lord Collingwood*) landed a record catch at Great Yarmouth on the 1st March 1947. The continued successful operation of many of the port's drifter/trawlers lasted right up until the 1960s. Steam and later diesel powered Great Yarmouth vessels rigged for trawling successfully worked the North Sea and also off the West Coast for a great many years. Historically, the drifter/trawler played a very important role at Lowestoft and Great Yarmouth, and in fact the very last trawl landing by a steam powered vessel in East Anglia was by one of these vessels. The drifter/trawler *LT235 Silver Seas* landed at Lowestoft on the 30th January 1960. As well as being the last steam trawling vessel to land, it was her last landing before being sold and converted to diesel power. The very last commercial steam trawler at Lowestoft, the 1902 built *LT154 Cairo* was laid up at that time, departing for the breakers yard on the 17th November 1959. After she had left, just one non-commercial steam trawler was to be found at the East Anglian ports.

By the mid 1930s, the first diesel powered trawlers had been built locally. After the Second World War an increasing number of these vessels progressively replaced the steam powered trawlers. Initially, the diesel trawlers were quite small side fishing vessels using similar fishing methods to the steam trawlers. In the 1960s, the stern trawler arrived, still using otter type trawls but fishing over the stern of the vessel. A major improvement in working conditions for the crew was instantly achieved by this method of fishing. The last large side fishing trawler, *LT30 Ripley Queen* was taken out of service and immediately cut up at Lowestoft in January 1994, after a life of 23 years. The withdrawal from service and replacement of trawlers using otter trawls was achieved in quite a short period of time. The relatively new stern trawlers had only a short fishing life, and were disposed of together with the side fishing trawlers. With years of life left in them, former Lowestoft stern trawlers can be found today in various parts of the world including South Africa, undertaking a variety of tasks. Many fishing companies ceased operations in the 1960s and 1970s. The last few trawlers using otter type trawls were succeeded by the more powerful and sophisticated beam trawlers. At one time three companies operated large beam trawlers from Lowestoft. The present fleet of modern beam trawlers at Lowestoft are all owned by the Colne Shipping Co. Ltd., one of the major leading progressive forces in the British fishing industry.

Section 3 - INSHORE FISHING BOATS

Great Yarmouth and Lowestoft both sustain a fleet of inshore fishing boats. As well as operating from each port of registration, considerable numbers of inshore fishing boats registered at Great Yarmouth and Lowestoft have, and still do, operate from nearby towns and villages. The majority of these boats operate from beaches. Lowestoft(LT) registered boats, have operated from such locations as Pakefield, Walberswick, Kessingland, Southwold (which has a harbour), and a few at Aldeburgh. Inshore boats may have been found at such locations as Winterton, Caister, Weybourne, Cromer, Wells(which has a harbour), and Sheringham with Great Yarmouth(YH) registrations. The continuing decline in the number of larger vessels has meant that the value of the inshore fleet to the industry has increased. A greater reliance has been put on these boats to ensure continuity of fish landings and supply. The inshore fleets over the years have undertaken drifting, trawling, long-lining and potting. Long-lining uses lines, hundreds of yards long with baited hooks attached. These are anchored on the sea bed and marked by buoys and flags. It is generally accepted that this form of fishing provides high quality fish, and tends to support selective conservation of fish stocks.

In addition to the inshore vessels, long-lining was also carried out by Lowestoft and Great Yarmouth drifters when it was economically viable.

SUMMARY

Lowestoft, could still be considered a major dual role fishing port in the late 1950's and early 1960's with well over one hundred trawlers engaged in white fish fishing and a herring drifter fleet of at least twenty five herring drifters(in 1958). With changing circumstances and the many problems that have beset the fishing industry, the Lowestoft fishing fleet has been reduced substantially with the total disappearance of side and stern trawlers and the herring drifters.

Mainly due to the perseverance, good housekeeping and forward planning by the directors and management of the Colne Shipping Co. Ltd., Lowestoft is in modern terms still a major fishing port.

Looking forward to the next millennium, the fishing industry in both Great Yarmouth and Lowestoft has reached a point of stabilised rationalisation, the latest major addition to the Lowestoft trawling fleet means that Lowestoft has the most modern "state of the art" trawler in Great Britain, whilst the inshore fleet at Great Yarmouth has seen in recent years new vessels added to the fleet. A substantial development at Lowestoft has been the regular visits by beam trawlers registered at other British fishing ports such as Grimsby, Buckie, Fleetwood, Peterhead, Hull, Ramsgate and Aberdeen. These visits seem set to continue and may increase with projected changes in European Common Fisheries Policy. The fishing industry has changed drastically and change has become a way of life. The threat of further regulation, is unfortunately ever present.

THE PHOTOGRAPHIC REVIEW

Towards the end of the 19th. century in Great Yarmouth harbour, the sailing drifter *YH229 Shade Of Evening* is seen with a sailing trawler moored alongside. *Shade of Evening* was built in September 1866 with dimensions of 48ft. x 15.2ft. x 6.5ft. She appears to have been laid up for some time when this scene was recorded. By the mid 1890s she had been broken up. On the trawler, one of the trawl heads on the beam is well demonstrated. It is of the type commonly used by Barking and Yarmouth sailing trawlers. Lowestoft and Ramsgate trawlers used a different pattern and Brixham trawlers another pattern. Visible on the trawl head is the eyebolt to which one of the bridles would be attached. When fishing with the beam trawl on the sea bed, these would be attached to the main trawl towing warp and thus the trawler.

All together - sail, steam and motor, trawlers and drifters. Early in the twentieth century, the Lowestoft paddle tug *Imperial* is seen towing two smacks, one of which is *LT93 Jessamine*, out of the harbour. Alongside an early Banff steam drifter heads for the fishing grounds together with a lugger rigged sailing drifter fitted with a motor. At the time, many of the motors fitted to these sailing drifters were built by Gardners. On the 12th. March 1917, the *Jessamine* was sunk by a German submarine 14 miles NNW of Trevose Head using a time bomb.

Approaching the entrance to Great Yarmouth harbour is the sailing trawler *YH902 Myosotis*. In the background an early steamship is seen heading south. *Myosotis* was built in 1877, she was 66.7ft long and weighed 66 gross tons. In 1894 she left Gt.Yarmouth after being sold to Plymouth and became *PH253*.

In this scene from the end of the 1880s, the sailing drifter *YH845 Boy Jim* is moored at Gorleston quayside. The previous King William the Fourth public house can be seen together with its lookout in the roof. Also visible is Moys coal depot. The *Boy Jim*, of 29.5 gross tons was built by Richards at Lowestoft in 1881. In February 1890 she was sold to Norway. Her last local owners were George and James Newson.

The Trawl Dock at Lowestoft before the 1893 extension was constructed and showing the area that would later become part of that extension. The work was authorised in October 1891, and involved the demolition of a number of buildings. The sailing trawler *LT245 Salus* seen on the left of the photograph, was built in 1889 at Galmpton and was sold to Sweden in 1912.

The new dock opened on the 1st October 1883 at Lowestoft was given the name of Waveney Dock. For several years it was referred to by many as the Herring Basin. This perhaps reflected the purpose for which it was built, to accommodate the seemingly ever increasing numbers of herring drifters using the port at that time in the autumn. In this very early photograph the large, rail served, herring and mackerel market can be seen. This market hall complete with Great Eastern Railway spandrels gave the impression of a large GER railway station. This magnificent building housed fish merchants offices and had a broad working area and quay.

A comprehensive photograph of Gt. Yarmouth harbour showing a large group of Smith's Dock Trust Co. Ltd. steam drifters, including *YH670 Twenty Two*, *YH691 Twenty Nine* and *YH674 Twenty Four*. Also present is the R.N.M.D.S.F. sailing trawler *LO340 Cholmondeley*, and various steam and sailing merchant ships. In the foreground is the beach yawl *Skylark*, run down by the steamship *F.E. Webb* in Yarmouth Roads in 1903 with the loss of three passengers, and three crew including Mr. Beckett and Mr. Shreeve. Of the drifters visible, *YH670* was sunk in June 1915 by a German submarine, *YH691* was sold to Wick in 1923 and scrapped in 1934 and *YH674* was lost on the 25th. March 1916 whilst on naval service.

Two views of Scotch fisher girls at work in the early part of the twentieth century. Many hundreds worked at Great Yarmouth, Gorleston and Lowestoft in the autumn months.

(Left) The Scottish fisher girls followed the drifter fleet in pursuit of the herring. Starting at Lerwick in the spring, and progressively moving south through Scotland, the North East of England and on to Great Yarmouth, Gorleston and Lowestoft in the autumn. Lowestoft was the furthest south these ladies would travel. Working in teams of three, they would gut, pickle and pack the herring. Two would be gutting and one packing. In the 1930s over 4000 were involved in this massive operation. The filling of barrels with brine is being carried out here by a young Scottish fisher girl.

(Right) Two of the crew of a sailing drifter involved in unloading their vessel at Lowestoft. The paddle tug in the background is the *Despatch,* belonging to the Great Eastern Railway. Built in 1875, she was lengthened in 1876, and rebuilt in 1907 with a single boiler and funnel. It was taken out of service in 1936, having been the main salvage tug at the port between 1875 and 1898. The new 105ft. screw tug *Lowestoft* arrived in 1898 and assumed the role of flagship of the Lowestoft tug fleet.

Introduced after the passing of the Cran Measure Act of 1908, the standard measure for herring is the cran. Prior to this, the amount of fish caught by a drifter had to be counted out by hand. Above can be seen the crew of a drifter engaged in counting out the fish which they had caught. The measures used at that time were: four herring made a warp, thirty-three warps made a long hundred (132 herring), and ten long hundreds equalled 1320 herring or fish, finally one hundred long hundreds made a last or 13,200 herring. The measures for mackerel were different, a long hundred being 120 mackerel, making a last 12,000 mackerel. The amount of fish equalling these various terms could however, vary according to geographic location and community. They also changed with time, the last initially being 10,000 fish . Another term was the cade, this eventually meant a quantity of 600 herring.

Reference to the Statute of Herrings Act of 1357, provides a detailed account of the terms and measures used prior to the cran.

The sailing drifter *LT661 Our Girls* leaving Lowestoft with the crew using sweeps to propel the vessel out of the harbour entrance. Locally the sailing drifters continued to be known as "luggers" in spite of the fact that since the late 1850s, the dipping lug sail had been generally replaced by a gaff ketch rig. Built in 1895, *Our Girls* was sold in 1903 to Lerwick.

The sailing drifters *YH42 Eclat*, *YH265 Venus* and *YH951 Paradox* being towed around the harbour bend at Yarmouth and out to sea by the paddle tug *Meteor*.
At both Great Yarmouth and Lowestoft the use of tugs to tow sailing fishing vessels in and out of the harbour was extensive, although not total. Some skippers preferred to take a cheaper option and have the crew row their vessel out of harbour using sweeps.
YH42 a 50ft.drifter of 25 tons was lost with all hands on the 10th. November 1891, *YH265* was built in 1862 at Gt. Yarmouth and was broken up in September 1898. *YH951* was built in 1884 by Richards at Lowestoft and broken up in 1909. The tug *Meteor* was built at North Shields in 1875. She was 94ft. long and of 97.64 gross tons. Her last local owner was the Great Yarmouth Star Steam Tug Co. Ltd. *Meteor* was broken up in August 1911.

A Kirkcaldy sailing drifter heading for the harbour entrance at Yarmouth around the turn of the century. The crew are using sweeps to assist in propelling the vessel. Nearer the entrance paddle tugs are towing sailing drifters to sea.

The paddle tug *Despatch* towing a group of sailing trawlers out of the harbour at Lowestoft early in the twentieth century. Also visible is the tug *Lowestoft*. The trawler nearest the camera is *LT244 Energy,* built in 1901 and sold to Sweden in 1912. Ahead of her is *LT325 Crystal*, built in 1904 at Lowestoft and sunk by a German submarine on the 27th January 1916, using a time bomb. With the replacement of sail by steam power in fishing vessels, the work of the railway tugs, was greatly reduced.

(Above Left) In the autumn of 1902 a Peterhead sailing drifter makes for the harbour entrance at Great Yarmouth with some of the crew using sweeps to help propel the vessel. The paddle tug *Tom Perry* is immediately astern of her, with at least six Scottish sailing drifters in tow heading for the sea. *Tom Perry* was built in 1879 and scrapped in 1920.

(Above Right) Early steam drifters moored in the Waveney Dock at Lowestoft. Nearest the camera are the Scottish drifters *INS 493 Venture*, built in 1902 at Lowestoft and *BF1411 Norseman*, built at Banff in 1903. Also visible are the Lowestoft drifter *LT1034 Silver King*, and the Wick drifter *Pansy*.

(Right) An everyday scene in the autumn of 1900 as a sailing drifter and sailing trawler are towed into Great Yarmouth harbour by a paddle tug. The trawler is *YH884 Boy Ben,* built in 1883 by Isaac Freeman Mack in Southtown. Her dimensions were 63.3ft x 17.1ft. x 8.3ft. Ahead of them, another paddle tug is towing several drifters and a trawler into port. *Boy Ben* was broken up in 1908.

Sam Richards yard at Lowestoft built their first steam drifter, *LT270 Test* in 1899. This was some years after other yards had started building steam powered fishing vessels. *Test* cost under £2000 to build for her Kessingland owners, C. and R. Harvey.

What is accepted as the first true motor powered drifter appeared in 1901 when Henry Reynolds built *LT 368 Pioneer*. This vessel was fitted with a 4 cylinder gasoline engine made by the Pennsylvania Iron Works Co., U.S.A. She was generally satisfactory in service and her performance silenced the majority of the many critics she initially had. *Pioneer* was sold in 1913, and left bound for Las Palmas under her own power.

The scene at the turn of the century on the South Denes, Great Yarmouth during the herring season.
The great majority of vessels visible in the harbour are sailing drifters with only a few steam vessels present.

The steam drifter *YH416 Lowestoft* was built in 1900 by Chambers and Colby at Lowestoft. She was 76.7ft. x 17.4ft. x 7.8ft., and had a net tonnage of 34.14 and a gross tonnage of 53.14. The wheelhouse placed aft was typical of many steam drifters and trawlers of that period. Her engine was a Elliott & Garrood compound of 15hp(130ihp). She was initially registered *H492* and owned by the British Coast Steam Fishing Co. Ltd. In 1916 she came under the ownership of R. Irvin, and became *YH416*. The *Lowestoft* was sold for scrap in 1923.

A superb view of the hull of an early steam drifter. The vessel is *LT164 Bounteous Sea* on the day of her launching. She was built by Chambers & Colby at Oulton Broad in 1900 and fitted with a 15hp(130ihp) compound engine manufactured by Elliott & Garrood. *Bounteous Sea* was sold for scrap in 1927.

Initially steam power was viewed with some apprehension by many owners of sailing vessels. However, these new vessels were soon to prove that they were capable of racing home to catch the best of the market, whilst the sailing vessels making the same journey, were becalmed or beating wearily against foul wind and tides.

A very large number of steam drifters were built in the early part of the twentieth century. *YH 567 Sunflower was* built at Beeching Bros. yard in Gt. Yarmouth in 1901. She was 78ft. 6ins. x 17ft. 6ins. x 8ft. 3ins. with a gross tonnage of 62.25 tons. *Sunflower* had a very short life. On the 24th. November 1904 she was totally destroyed by fire. She is seen here leaving the harbour mouth on trials in 1901.

The steam drifter *LT269 Daily Bread* was built in 1901, as yard No.85 at the Lowestoft shipyard of Sam Richards. She was built for Mr. J.C.Cooper. Later he was to sell the vessel to Mr. J. E. Capps. The drifter is seen here leaving Lowestoft on her way to the fishing grounds. In 1920 the *Daily Bread* was sold to the Admiralty.

The steam trawler *LT1031 Alaska* was built of iron in 1898 at at North Shields and previously registered *GY477*.
She is seen here approaching the harbour entrance at Lowestoft.
In 1926 *Alaska* was sold to France.

The sailing trawler *YH1039 Better Luck* was built in 1885 by E. Castle & Sons at Southtown. She was later registered *LT418* and also *LT185*. As well as working as a trawler, in the home fishing she worked as a fish carrier to the Continent. In 1908 she also made trips to the Shetlands with new nets, returning with spoiled nets and livestock.
In 1913 *Better Luck* was sold to Sweden.

(Above) A view of Beeching Bros. yard at Gt.Yarmouth in 1902 with a number of sail and steam fishing vessels under construction. Furthest right is the Gt.Yarmouth steam drifter *YH703 Yare*. She was built for Great Yarmouth Steam Drifters Ltd. and had a Crabtree compound engine. *Yare* was 79ft. long with a gross tonnage of 57.62 tons. In 1914 she was sold to Cullen and became *BF99*. *Yare* was scrapped in 1923.

(Right) Launch day for the Gt.Yarmouth steam drifter *YH359 Harry*. Built at her home port in 1901 by Fellows, her initial owner was Herbert J. Sayers. Her dimensions were 74ft. 2ins. x 18ft. 2ins. x 8ft. 3ins. In 1913 she passed into Scottish ownership and became *FR527*. *Harry* was lost in the North Sea on the 9th. August 1927.

Later to be run down on the 2nd October 1924, by the steam trawler *LT427 Lolist*, the steam drifter *LT471 Pride* is seen moored in front of the thatched ice house at Lowestoft. The passengers and crew have been assembled for the photographer, shortly before leaving on a trip to sea, on completion of the vessel in 1904. *Pride* was built at the John Chambers yard at Oulton Broad. The ice house was used to store ice brought from Scandinavia by sailing ship prior to the building of the ice factory.

Built in 1904 at Beverley by Cook, Welton and Gemmell the 78ft. Yarmouth steam drifter *YH 897 The Princess* was owned by The Crown Steam Drifters Ltd. She was fitted with a Crabtree compound engine and was of 77.49 gross tons. In 1924 *The Princess* was sold to Banff and became *BF40*. She was sold for scrapping in 1937. In this view *The Princess* is seen with the Shetland Islands in the distance.

Crew members of the sailing trawler *LT517 Prairie Flower* line up on deck in the Inner Harbour at Lowestoft. The *Prairie Flower* was launched with a bottle of water at Oulton Broad in 1892. Sold to Sweden in 1911, she was scuttled in 1975.

A rare photograph of at least 30 sailing trawlers becalmed off the coast in 1904.

(Above left) Built at Lowestoft in 1905, the steam drifter *LT545 Livonia* is seen approaching the pier heads at Lowestoft on her way to the fishing grounds on the 23rd October 1933. Her drift nets and buffs are clearly visible on deck.

(Above right) Demonstrating one of the lesser known skills associated with drift net fishing are these cork cutters. In some cases the registration of the drifter would be marked on the corks and on the lead colletts.

(Bottom right) The steam drifter *YH291 Young Archie* seen here entering Great Yarmouth harbour, was built in 1908 at Lowestoft by John Chambers. Powered by a Elliott and Garrood compound engine she was of 59.5 gross tons and 74ft. 6ins. in length. In 1930 *Young Archie* was sold and became *BK24*. She foundered in the North Sea in February 1935.

Leaving Lowestoft is the R.N.M.D.S.F hospital ship *LO51 Queen Alexandra*. Built at Leith in 1902, she was sold for scrapping in 1934. This scene was recorded on the 6th. August 1931.

On the 12th. October 1935, *LT 445 Auckland* is seen leaving Lowestoft for another fishing trip. *Auckland* was formerly registered *H441* and like a great many drifters and trawlers, she was engaged in minesweeping during the First World War. Constructed of iron in 1899 at Hull for the Hull Steam Fishing & Ice Co. Ltd., she was sold for scrapping in 1937. *Auckland* was powered by a 35hp triple expansion engine made by Amos and Smith.

Mission ships owned by the R.N.M.D.S.F. were a familiar sight in Great Yarmouth harbour for many years. Moored in the harbour on the 10th. July 1930, was *LO51 Queen Alexandra*.

In 1907 Harry Reynolds at Oulton Broad built the motor powered drifter *LT1035 Thankful*. Her propelling machinery was made locally by Brauer & Betts. The internal combustion engine had two cylinders and developed 60ihp at 280/300rpm. The vessel was reasonably successful. However, in 1908 the motor was removed and a compound steam engine built by Elliott & Garrood installed, together with a boiler made by Dodman's. *Thankful* was sold in March 1918 to Whitby and became *WY241*. In 1925 she was sold for scrap. *Thankful* is seen in the photograph in her motor powered days.

Two further examples of early steam drifters were *LT463 Reward* and *LT1017 Briton,* both built at the Lowestoft shipyard of Sam Richards in 1906.

LT463 Reward sets out on sea trials from Lowestoft with a full complement of invited guests. *Reward* was built for R. and J. G. Utting of Kessingland.
On the 5th. December 1929 she sank in the roads off Great Yarmouth.

LT1017 Briton cost £2,370 to build. From the placing of the order to handover took 40 working days, a severe penalty clause being included in the order in the event of late delivery.
Cyril Richards, a son of the founder of the shipyard, was heard to say that when the order was placed, her keel was growing on a tree near Beccles.
He later travelled on her from Lowestoft to Penzance and recorded the journey time as 44 hours. Unusually on completion, this vessel was finished in French Grey. She was sold for scrap in 1936.

The Lowestoft smack *LT1203 Masterhand* at anchor with two other smacks. The *Masterhand* was one of the last sailing trawlers to remain in service at Lowestoft in the late 1930s. She was built in 1920 at Rye by G. and T. Smith & Co. and was sold away from Lowestoft in 1946. She was re-registered as *BM43* after being fitted with an engine, and continued to fish until the late 1960s.
The remains of the *Masterhand* can be seen on the River Tamar.

The Lowestoft smack *LT404 Greyhound* underway in Scottish Waters in the 1920s. Built in 1903 at Lowestoft, this 68ft. vessel was sold to Norway in 1931.

The "Strath." class steam trawler *LO319 William Harrison* seen leaving Lowestoft in 1920. Built at Wivenhoe in 1919, *William Harrison* was sold in 1921 and became *GY1335 Flavia*. In 1935 she was sold again, moving to Aberdeen and given the new registration *A373*. She was lost on the 28th. August 1940.

40

The 62ft. sailing trawler *LT394 Godetia* was built at Galmpton in 1909. For the early years of her life she was registered *R261,* but in 1919, *Godetia* became *LT394.* In 1939 she was sold and made the voyage to her new home in Norway.
On the opposite page, top left, the *Godetia* is seen leaving the Trawl Dock for sea.

The other two photographs opposite, and those on this page are rare authentic photographs taken on board the *Godetia*. Three of these are taken at sea. In the photograph on the opposite page bottom left, the crew are shooting the beam trawl. The other photograph is a deck scene with the vessel sailing. On this page, above, a poor but valuable photograph shows some of the crew gutting while the vessel is still trawling. The view on the right, shows the vessel back at Lowestoft with the landing of the catch taking place. Amongst the many features visible on these photographs of this working smack, are the steam capstan, boiler chimney, towing post, thole pin, main trawl warp and the tiller.

These three steam drifters were all built at Lowestoft for Great Yarmouth owners.

(Above left) *YH 766 Kiama* was completed in 1914. Built by Richards and fitted with a Richards 30hp(145ihp) compound engine. She was sold in 1919 and became *LT205*, in 1928 she was sold again became *YH359 Girl Ellen*.

(Above right) YH22 *Eastholme* was completed in 1912. Built by John Chambers and fitted with a Elliott and Garrood 20hp(130ihp) compound engine, she was run down and sank 35miles from Flamborough Head on 9th. September 1934.

(Bottom right) *YH 627 George Albert* was completed in 1916. Built by Richards and fitted with a Burrell 42hp compound engine, she was named *G.A.W.* until 1927. Sold for scrapping in 1947.

42

With two sailing trawlers in the distance and the compass adjuster on top of the wheelhouse, the steam drifter/line fishing vessel *LT451 Flo Johnson* leaves Lowestoft. Built in 1917 at Oulton Broad, she was registered *SH325* until 1919. In 1927 *Flo Johnson* was sold to the Star Drift Fishing Co. Ltd., and became *Abiding Star*. She was sold for scrapping in 1947.

The scene at Great Yarmouth during the home fishing in the 1920s. Scottish steam drifters form the majority of the vessels in this scene. Two local drifters are identifiable, *YH734 Moss* and *YH350 Pleasants*.

A prominent feature of the Great Yarmouth fishing industry, were the swills, the baskets seen in both these views and elsewhere in this book.

The 86ft. *YH734 Moss* was part of the Westmacott Ltd. fleet for many years. She was built in 1911 at Mackie & Thompson's yard in Govan. In 1932 *Moss* was sold to Peterhead and became *PD172 Guiding Light*. During 1953 she was sold for scrapping.

YH350 Pleasants was built by Fellows in 1913 and fitted with a F.W. Carver 24hp compound engine. Her boiler was made by T.Sudron of Stockton-on-Tees and had a working pressure of 140lb./sq. inch. She was sold in 1927 and became *LT340 John Alfred*. The *John Alfred* was sold for scrapping in 1946 at Oulton Broad.

During the First World War, plans for the Admiralty " standard " drifter were drawn up. The naval roles for these vessels was undertaking duties such as minesweeping, acting as tenders and examination vessels. These vessels were built at different shipyards around the country to a standard specification. Two of the shipyards to build these drifters were those of J.W. Brooke and Colby Bros. at Oulton Broad. After the war the majority of these drifters were sold to fishing vessel owners. (Main photograph) *H.M.D.Flash* was built at Colby Bros. in 1918. It is seen on launch day and was built of steel, not as indicated on the photograph. *Flash* became *FD379 A.J.A* and later *GY309*, it was sold to Spain in 1924. (Top photograph) Many of the vessels such as *LT 564 Blue Haze*, seen here at Lowestoft on 11.9.27, were used as drifter/trawlers. Built by J.W.Brooke, it later became *PD67 Lilium*. It was sold for scrapping in 1956.

Consolidated Steam Fishing & Ice Co.(Grimsby) Ltd. owned many wheelhouse aft steam trawlers early in the twentieth century. Three of their trawlers are seen here at Lowestoft. At one time all three had been Grimsby registered.

(Above left) Built in 1897 at Govan, *LT 134 Newhaven* was until 1925 *GY232*. She was transferred to Lowestoft in 1920. *Newhaven* had just over two years to live when photographed on the 26th. October 1937. On the 15th. January 1940 she was mined 18 miles SSE of Lowestoft.

(Above right) Built in 1900 of iron at Selby, *LT248 Dereham* was formerly *GY1146 Rinto*, owned by G. F. Sleights. She left Lowestoft on the 4th May 1954 for scrapping at Southampton.

(Right) Seen leaving for the fishing grounds on the 28th. September 1928, is *LT355 King Henry*. She was formerly *GY1169*. On the 13th. June 1941 *King Henry* was sunk during one of the many German air attacks on Lowestoft. She was built in 1900 at Grimsby.

The repair of herring nets was undertaken by hundreds of beatsters working either individually at home, or in teams in net stores owned by the fishing vessel owners. Many of these net stores are to be seen today, converted for numerous uses in Gorleston, Lowestoft, Great Yarmouth, and some surrounding villages such as Kessingland. In this photograph beatsters are working in a Great Yarmouth net store.

One of the hundreds of Scottish steam herring drifters that visited Gt. Yarmouth and Lowestoft for the home fishing. *BF 198 Rowan Tree* was built at Sandhaven in 1918. She is seen here entering Lowestoft harbour. Completed as an Admiralty drifter, she capsized off Lowestoft on the 21st. November 1941.

Many of the local drifters worked in the northern North Sea in pursuit of the herring. Ports such as Lerwick, Peterhead and Scarborough were regularly visited by Lowestoft and Yarmouth drifters.

(Right) The Lowestoft drifter *LT1171 Rewold* is seen in Lerwick. *Rewald* was built in 1912, sold in May 1939 to become a yacht, and scrapped in 1953.
(Bottom left) Scarborough harbour in the 1920s with the Lowestoft steam drifter *LT61 Dick Whittington* unloading.
(Bottom right) Peterhead with the Yarmouth steam drifter *YH252 Jack George* leaving port with a Peterhead steam drifter following. The 89ft. *Jack George* was built by Sam Richards at Lowestoft in 1913. She was owned by various members of the George family of Winterton during her life. Her engine was a 25hp(160ihp) triple expansion steam engine by Elliott & Garrood. *Jack George* was sold for scrapping in 1948 to Seago's of Great Yarmouth.

Views of Lowestoft and Great Yarmouth harbours during the home fishing in the 1910s and 1920s.

In this scene, Scottish steam and motor drifters are leaving Great Yarmouth for the fishing grounds.

Local drifters packed into the Waveney Dock at Lowestoft unloading their catches. No motor vehicles can be seen, all the transport being horse drawn.

Three further examples of steam trawlers from the fleet of Consolidated Steam Fishing & Ice Co. Ltd., later to become Consolidated Fisheries Ltd.

(Above) A vessel which had five different registrations during her life was *LT84 Croton*. This famous trawler was built in 1898 at Govan and left for the breakers yard on the 2nd. Feb. 1950. During her life she worked for Boston Deep Sea Fishing & Ice Co. Ltd., Consolidated Fisheries and J. Marr & Son Ltd. *Croton* was powered by a 41hp 3cyl. Russ & Duncan engine and was 100ft. in length and of 150 gross tons.

(Top right) *LT319 Loddon* was built in 1919 at Bowling. After many years service at Lowestoft she was sold to Aberdeen and became registered *A8*. *Loddon* was sold for scrapping in 1960.

(Bottom right) Seen leaving Lowestoft on the 4th. October 1928, is *LT160 King Athelstan*. She was formerly *GY97* and *GY258 King Egbert*. *King Athelstan* was built in 1899 at Grimsby and sold for scrapping in 1953.

50

Other typical steam trawlers of Lowestoft.

(Below) Seen leaving Lowestoft is Heward Trawlers Ltd., steam trawler *LO200 Junco*. Built in 1917, she was sold for scrap in 1957. She was one of last of the hundreds of "Short Blue" trawlers in East Anglia.

(Top right) *LT977 Rosalind* was built in 1905 at Beverley, and. was initially registered *H839*. She was powered by a 45hp Amos & Smith engine. *Rosalind* was sold for scrap in 1955. On the 3rd. October 1933, she is seen heading out on another fishing trip.

(Bottom right) The trawler *LT405 Owl* was also registered *GY44, FR249* and *A618* during her life. This photograph of her was taken on 7th. October 1929. *Owl* was built in 1896 by Edward Bros., North Shields, and fitted with a North Shields built engine. She was sold for scrapping in 1935.

Both wooden steam drifters on this page were built at Lowestoft, and fitted with Elliott and Garrood engines.

YH176 Jacob George was built in 1910 and fitted with a compound engine. During her life she also undertook seine netting. *Jacob George* was sold for scrapping in 1949.

YH264 Ocean Spray was fitted with a triple expansion engine when built as yard No.179, at Sam Richards shipyard for Mr. S. George. She was sold for scrapping in 1957.

Scenes from the 1920s.

(Left) Steam drifters from Fraserburgh and Kirkcaldy leaving Great Yarmouth harbour. It would appear that they are very close to having a collision.

(Bottom Left) Steamers being loaded with barrels of herring for export at Lowestoft North Quay.

(Bottom right) A sight never to be seen again, as scores of steam drifters converge on the entrance to Great Yarmouth harbour. As far as can ascertained these are all Scottish vessels.

The steam drifter *YH831 Holly* was one of the famous Westmacott Ltd. fleet for many years. In 1932 she was sold to Lowestoft drifter owners P.W. Watson & Sons Ltd.

Built in 1904 at North Shields by Smith's Dock Co. Ltd., the *Holly* is seen in this early photograph leaving her home port for the fishing grounds. The famous Smith's Dock Co. at Middlesborough closed in 1987. It built fishing vessels in addition to the many large vessels and rigs it was renowned for. It was the last of many shipyards on the Tees. Their yard at North Shields where *Holly* was built, completed its last sea going ship, the *Mountcharles* in 1909. The Smith Dock Co. had a substantial interest in the fishing industry at Gt. Yarmouth through the Trust Co. it formed. The *Holly* was of a similar design to vessels of the Smith's Dock Trust Co. Ltd fleet at Gt. Yarmouth.

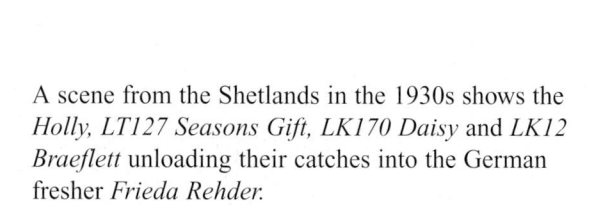

A scene from the Shetlands in the 1930s shows the *Holly*, *LT127 Seasons Gift*, *LK170 Daisy* and *LK12 Braeflett* unloading their catches into the German fresher *Frieda Rehder*.

An early casualty of the Second World War whilst allocated to H.M.S. Vernon, and engaged in mine trawling activities was *LT230 Ray Of Hope.* On the 10th. December 1939 she was mined with serious loss of life in the Thames Estuary. *Ray Of Hope* was built by J. Chambers at Oulton Broad in 1925. The date of this photograph is the 6th. August 1930, when she was off Lowestoft.

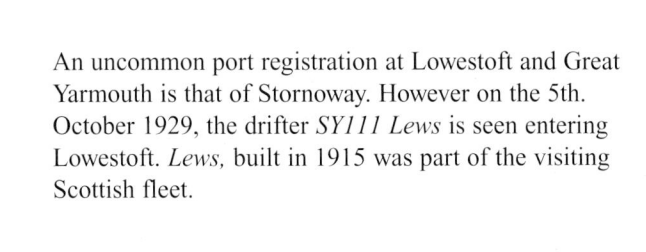

An uncommon port registration at Lowestoft and Great Yarmouth is that of Stornoway. However on the 5th. October 1929, the drifter *SY111 Lews* is seen entering Lowestoft. *Lews,* built in 1915 was part of the visiting Scottish fleet.

For many years during the autumn home fishing, Scottish steam and motor powered drifters worked side by side. The last Scottish steam drifter visited East Anglia in 1957.

A typically very smart Scottish steam drifter seen leaving Lowestoft, *BF107 Walkerdale* was built in 1911 at Govan. In 1946 she became *BCK35 Hazelgrove,* and in 1950 she was converted to diesel power. For many years *Hazelgrove* was owned by members of the Wilson family at Fleetwood. *Hazelgrove* was scrapped at Fleetwood in 1968.

The motor drifter *BF1091 Laurel* was owned by B. & J. Sutherland of Findochty. The vessel was powered by a 75hp Gardner engine and is seen here on the 4th October 1935, entering Lowestoft.

The 1920s saw a number of Aberdeen registered steam trawlers bought by Lowestoft owners. Two of these large vessels are featured here. After a few years they were resold to Aberdeen due to operational difficulties at Lowestoft. Both vessels are seen in Lowestoft harbour.

LT 511 Strathlossie was built in 1910 at Aberdeen. Her previous registration was *A316*, and on returning to Aberdeen she became *A952*.

Prior to coming to Lowestoft, *LT532 Strathderry* had the registration *A401*. In 1927, on return to Aberdeen she became *A226*. She was sold for scrapping in 1955.

LT10 Ocean Scout was bought in 1947 by a salvage company and later was left on the foreshore of Lake Lothing, Lowestoft. Built in 1913 for Chapmans Steam Drifter Co. Ltd., she was fitted with a 20hp. Elliott & Garrood triple expansion engine.
She is seen here in 1926 approaching her home port.

Many will know the name of *Togo*, the vessel which was one of the first successful fishing vessel conversions from steam to diesel power. Built in 1905 at Great Yarmouth, she is seen here as the steam drifter *YH248,* before conversion. The conversion was carried out by L.B.S. Engineering between October 1934 and March 1935, after the vessel had been sold. The diesel engine installed in *Togo* was a 3cyl. 200hp. Mirrlees. For a time *Togo* was registered as *LT609*, and eventually she became *LT69*. On the 6th October 1964 she left for the shipbreakers yard towing the much larger trawler *Tobago,* also to be scrapped.

Built in 1915 at Oulton Broad, *YH312 Ocean Toiler* undertook drifting, lining and seine netting. She is seen entering Lowestoft on the 5th. September 1926. *Ocean Toiler* was a member of the famous Bloomfields Ltd. fleet for most of her life. She was sold for scrapping in 1949.

Heading out of Great Yarmouth on the 21st. October 1932 for the fishing grounds is *YH74 Supporter*. She and was built in 1914 by Colby Bros. at Oulton Broad and formerly registered *LT119*. Her triple expansion engine of 20hp(160ihp) was built by Elliott & Garrood at their Beccles works. *Supporter* was a total loss on the 2nd. November 1944.

A scene from the Shetland Islands in the 1930s. Many Yarmouth and Lowestoft drifters are present together with Scottish drifters. The more prominent vessels are *YH172, YH359, YH519, LT44, LT200, LT256, LT655 and LT777*.

The Lowestoft wooden steam drifter *LT342 Eileen Emma* was built in 1914 at John Chambers yard in Oulton Broad. In 1915 she saved 116 survivors from the torpedoed S.S. *Falaba*. *Eileen Emma* is seen here in 1932 in Scottish waters. In 1946 she was sold to Norway and became the *Farroy*.

By the late 1920s and early 1930s, many sailing trawlers had successfully had diesel engines installed in them by the Sam Richards shipyard at Lowestoft. Such was the success of these installations, an order was placed in 1931 by W. H. Podd Ltd. for a twin screw diesel drifter/trawler. This was to become *LT245 J.A.P.*, and she is featured elsewhere in this book. In 1932 this shipyard became Richards Ironworks Ltd., after receiving an order for twelve 75ft. steel trawlers. Each was powered by a 150hp Ruston diesel engine. Because of the Second World War only seven were built. One of this pioneering class, *LT57 Eta* is seen above, hauling her gear. During the Second World War in January 1940, this vessel was sunk by mine damage.

The major success of the small diesel trawlers completed by Richards between 1933 and 1935, brought an order in 1935 for a larger diesel trawler from Grimsby Motor Trawlers Ltd., Grimsby. This vessel was *GY153 British Columbia.* She was powered by a 310hp Ruston diesel engine. It was to be the first successful diesel trawler at that port. This 101ft. vessel was later sold to the Clan Steam Fishing Co.(Grimsby) Ltd., and operated from Lowestoft with the registration of *LT404.* In September 1957, the *British Columbia* was lost in the North Sea after a collision with an American destroyer. She is seen in this photograph at her builder's fitting out quay.

(Top left) The Lowestoft trawler *LT96 Yulan* was formerly *GY348*. She was 96ft. in length, of 143 gross tons and fitted with a 45hp. Amos & Smith triple expansion engine. *Yulan* was sold for scrap in 1948.

(Bottom left) Seen leaving Lowestoft on the 26th. September 1929 with her Lowestoft registration of *LT396* is *Bellerophen*. From 1907 when she was built at Selby, until 1928, *Bellerophen* was registered as *GY335*. In 1954, *Bellerophen* was sold for scrap by Walton Fishing Co. Ltd. She was 88 tons nett and 184 tons gross, with dimensions of 105ft. x 21ft. 6ins. x 11ft. 1ins. She was powered by a 3cyl. 57hp triple expansion engine by Holmes.

Both *Yulan* and *Bellerophen* were owned for many years by Consolidated Steam Fishing & Ice Co.(Grimsby) Ltd., Grimsby.

(Top right) *YH33 Ocean Retriever*, seen here entering the harbour at Great Yarmouth, was built in 1914 by John Chambers Ltd., at Oulton Broad. On the 4th. August 1931, she became a total loss on rocks near Mallaig harbour. *Ocean Retriever* was owned by Bloomfields Ltd.

The Lowestoft drifter *LT125 Homeland* off the Shetlands in the 1930s. She was built in 1908 by Richards at Lowestoft for Mr. A. Jenner. During the First World War, she landed one survivor from the submarine *D5* which had been mined. *Homeland* had a gross tonnage of 82 tons with dimensions of 84ft. 7ins. x 18ft. 6ins. x 9ft. 2ins. She was sold for scrapping in 1938.

One of the many smacks fitted with diesel engines in the 1920s and 1930s was *LT73 Rosary*, built in 1924 by C. & T. Smith at Rye. She is seen here on trials, after completion of the installation of her 125hp. 5 cyl. Crossley diesel engine in 1934. She was sold to Aberdeen in 1941 and became *A520*. *Rosary* was sold for scrapping in 1952.

A busy scene at Lerwick in the 1920s, the Lowestoft drifters *LT 1296 Landbreeze* and *LT1297 Olivae* can be seen, with the unloading of the *Landbreeze* in progress. The women are gutting at the farlins as a fully loaded bogey is pushed along the pier railway. This operation of unloading the drifters, and processing the fish was well planned and efficient. *Landbreeze* was built in 1919 as a standard drifter and sold for scrapping in July 1955. *Olivae* was built at Gt.Yarmouth in 1915, and sold for scrapping in April 1956.

Leaving harbour on trials in 1933 after being fitted with an engine by Richards Ironworks Ltd, is the former smack *LT249 Purple Heather*. The engine was a 100hp 4cyl. Crossley diesel. In 1943 she was sold to North Shields and later she was sold to Denmark. The vessel was wrecked in 1952. Built in 1921 by John Chambers Ltd. at Oulton Broad, this was the second vessel to have this registration and name. The first smack with this registration and name was destroyed by a time bomb, placed on board by a boarding party from a German submarine on 12th July 1915.

On the 8th October 1937 the Lowestoft drifter *LT519 Ambitious* was involved in a collision with *BF196 Spectrum* off the piers at Lowestoft. This resulted in the vessel becoming holed. In the photograph above, she is settling in the water. *Ambitious* was built in 1910 at Richards shipyard, Lowestoft.

The vessels used in raising the *Ambitious,* were the Lowestoft tug *Lowestoft* of 1898, the Yarmouth tug *George Jewson* of 1908, the dredger *Grabber* of 1931, and the dredger *Pioneer* of 1886. The *Ambitious* was finally raised on the 19th. October 1937. She was later broken up.

In the photograph on the left, can be seen the dredger *Grabber*, the *Ambitious* and the tug *Lowestoft*. In the photograph on the right is the *Ambitious*, the tug *George Jewson* and the dredger *Pioneer*. In both photographs, local and Scottish drifters are seen entering and leaving the harbour.

Looking astern from a drifter/trawler returning to port in the 1930s. The Lowestoft drifter *LT777 Lord Zetland* is following, on her way back to port from the fishing grounds. Two drifters are visible heading for the fishing grounds. *Lord Zetland* was built in 1914 at Lowestoft and for many years was owned by the Lowestoft Steam Herring Drifter Co. Ltd. She was sold for scrap in 1951.

Unlike most fishing ports, until the late 1930s, a common sight at Lowestoft was sailing, steam and diesel trawlers, plus steam drifters all sharing the same facilities.

(Right) About to enter the Waveney Dock in the 1930s is *LT 427 Lolist*. Built in 1914 at Middlesborough, she was sold in 1949 to Scotland.

(Below) The last of the Short Blue sailing trawlers *LO392 Boy Leslie* seen leaving Lowestoft on the 8th. August 1935. In May 1939 she was sold to new owners in Norway.

At sea on a drifter in the 1930s, the Lowestoft drifter *LT304 Sunbeam II* is passing on her way to the fishing grounds. *Sunbeam II* was built in 1916 at Oulton Broad, and was previously registered *YH279*. She was owned at Lowestoft by the well known Pakefield drifter owner Mr. J. J. Colby, and was sold for scrapping in 1954.

This view, complete with steam roller on the quay, shows *YH997 Girl Winifred* moored in the river at Gt. Yarmouth. It was taken on the 5th. August 1931, and shows many aspects of the area which have long since disappeared. The 86ft. *Girl Winifred* was built in 1912 at Gt.Yarmouth by Fellows. She had a Burrell 25hp(100ihp) compound engine and Riley Bros. boiler. The *Girl Winifred* was sold for scrapping in 1951.

Lost on the 18th. April 1941 off the River Tyne, whilst on naval service, *YH55 Young Ernie* was one of the first Yarmouth vessels to be fitted with wireless. Built at Lowestoft by John Chambers in 1925, she is seen here at the harbour entrance at Gt.Yarmouth leaving for the fishing grounds. Her dimensions were 85ft. 8ins. x 19ft. 7ins. x 9ft. 8ins. She was fitted with a Elliott & Garrood triple expansion engine and was of 88.1 gross tons.

The trawler *LT159 Valeria* was registered *GY818* when she arrived at Lowestoft on the 1st. April 1924. She was built in 1898 at Beverley of iron and powered by a Holmes 55hp triple expansion engine. *Valeria* became *LT156* in 1925. This photograph was taken on the 26th. August 1930, and shows her leaving Lowestoft. On the 18th. August 1940, she was attacked by German aircraft and sank 8 miles from The Smalls.

On the fishing grounds and hanging onto the nets. This view shows the drifter *LT1120 Present Help*, drifting with her nets in the 1930s. She was built by Crabtree at Gt.Yarmouth in 1911, and was 84.6ft in length and 81 gross tons. She last fished in 1952 and was sold for scrap in May 1953.

Aspects of Lowestoft Harbour in the 1930s

(Top) Inside the Sale Ring, Waveney Dock.
(Bottom) Consolidated Fisheries vessels laid up in the Inner Harbour.

(Top) Trawlers in the Trawl Dock, with a coal barge entering.
(Bottom) Drifters laid up in Hamilton Dock

The Yarmouth drifter/trawler *YH29 Ocean Lifebuoy*, a vessel of the Bloomfields fleet, entering Gt. Yarmouth harbour on the 14th. October 1931. Built by Alexander Hall & Co., Aberdeen in 1929, she was sold in 1955 to W.H.Kerr (Ship Chandlers) Ltd. In this photograph she is rigged for drifting. Elsewhere in this book, she is rigged for trawling. She was powered by a 3 cyl. 39hp triple expansion engine built by her builders. Under her new owner, *Ocean Lifebuoy* was renamed and converted to diesel power.

A Peterhead steam drifter leaving Great Yarmouth in the early 1930s during the home fishing.

The wooden steam drifter/trawler *LT288 Renascent* represented a class of vessel rather uncommon at Lowestoft. Built by Fellows at Great Yarmouth in 1926, she was sold to Norway in 1946. On the 28th. October 1946, the *Renascent* foundered while on passage to her new owners. This rare photograph shows her on the 8th. October 1934, rigged for trawling and leaving Lowestoft.

This fine example of an early steam trawler was sunk during an attack on Lowestoft by German aircraft on the 13th. June 1941. We see her here at sea off Lowestoft. *LT355 King Henry* was built in 1900 at Grimsby. This 105ft. vessel was owned by Consolidated Fisheries Ltd., Grimsby and until 1927 had the registration *GY1169*.

YH7 Boy Ray was owned by G. R. Newson and until 1934 was *LT736 Rajah of Mandi*. She was built in Great Yarmouth in 1910. *Boy Ray* is seen on the 26th. October 1938, at Gt.Yarmouth returning from a fishing trip.

Drifters heading for the harbour entrance at Great Yarmouth with *YH464 Girl Ena* leading. Astern of her is *YH217 Frons Olivae*, two Lowestoft drifters and a Kirkcaldy drifter. Other drifters are following down the river. *Girl Ena* was built in 1907 by Fellows at Gt.Yarmouth, and was initially registered *LT3*. She had a number of owners, her last local owner being Norford Suffling Ltd. In 1947 she came under Scottish ownership, and in 1949 was sold for scrapping.

The hull of the steam drifter/trawler *LT122 Feasible* can still be seen today. She is seen here entering Gt. Yarmouth harbour in the 1930s. *Feasible* was built in 1912 at Aberdeen, and fitted at Lowestoft with a 25hp Elliott & Garrood triple expansion engine. Initially registered *LT1191* she was sold in 1919 and became *R157*. She returned to Lowestoft in 1930 and became *LT122*. Sold to Norway in 1946, she was converted to a motor coaster. In this form *Feasible* can be seen today, now owned by a preservation society.

Maintenance of fishing gear is ongoing. In this scene from the 1930s, drift net buffs are being painted and hung out to dry.

No example of the wooden steam drifter has been preserved. These vessels made a major contribution to the British fishing industry, and it is only to be expected that a large number would appear in this book. A number of these sturdy vessels still exist in various conditions around the world. At Lowestoft, the remains of one can be seen, still containing her engine.

Laid down as the Admiralty drifter *HMD Moonlight*, but completed privately, this superb photograph shows *LT553 Carry On* at the entrance to Lowestoft harbour. Completed in 1919, she was mined on the 17th. December 1940 off Sheerness. This fine classic scene was recorded on the 6th. April 1931.

The one and only *LT100 Formidable,* made famous by the late Ted Frost in his book "From Tree to Sea". This book explains in great detail the building of a wooden steam drifter and in particular *Formidable*. She was built in 1917 at John Chambers yard where Ted worked. Her details are given as 88ft. overall, with a beam of 19ft. and a depth of 9ft. 9ins. She was fitted with a Riley boiler and Crabtree engine. The Elliott & Garrood steam capstan was No. 5947 and she was designed to carry 300 crans of herring, weighing about 52 tons. *Formidable* would run economically at a speed of 9 knots. She was sold in 1946 to Norway. She is seen here crossing Waveney Dock on the 20th. August 1937.

Vessels of the famous Bloomfields fleet of Great Yarmouth have traditionally been regular visitors to Lowestoft. In these scenes, two sisterships of their fleet, one rigged for trawling, the other drifting, are seen at Lowestoft. Both were built in 1929 at Aberdeen.

With a number of drifters following, *YH28 Ocean Sunlight* heads into Lowestoft on the 10th. October 1930. She was mined on the 13th. June 1940 off Newhaven.

A few days prior to the scene on page 74 where *YH29 Ocean Lifebuoy* is a drifter, we see her here on the 2nd. October 1931, as a trawler. She was sold in 1955 by Bloomfields Ltd., and renamed *Deelite*. During 1958, *Deelite* was converted to diesel power at Lowestoft by Richards Ironworks Ltd., for her owners W. H. Kerr (Ship Chandlers) Ltd. of Milford Haven. The dimensions of *Ocean Lifebuoy* were 94ft. x 20ft. 1ins. x 9ft. 7ins., as a steam vessel she had a gross tonnage of 131 tons. She was sold for scrapping in 1973.

The Lowestoft drifter/trawler *LT344 Lord Anson* was for many years a member of the Lowestoft Steam Herring Drifters Co. Ltd. fleet. She was sold away from the port after the Second World War and spent many years working on the West Coast. Built in 1927 at Selby, *Lord Anson* was powered by a 42hp Pertwee & Back triple expansion engine. She was sold for scrapping in 1957.

Looking at this 20th. October 1933 scene, it is difficult to believe that three years later, *LT929 Encore* would be sold for scrapping. She was 105ft. long, formerly registered *H523*, and had a 40hp Amos & Smith engine. *Encore* was built in 1900 at Hull.

The entrance to the harbour at Great Yarmouth as seen from an approaching vessel. A Fraserburgh steam drifter is clearing the entrance and heading for the fishing grounds.

The well known Haylett family drifter, *YH622 Animate,* was built at John Chambers Oulton Broad shipyard in 1917. She is here about to clear the entrance of Yarmouth harbour and head for the fishing grounds. *Animate* was requisitioned for naval service in both wars and sold for scrapping at Peterhead in 1957. She was of 74.53 gross tons and powered by a F.W. Carver 24hp(150ihp)compound engine.

A group of Lowestoft drifters together with Scottish vessels in the Shetlands in the 1930s. Visible are *LT76 Excel*, *LT238 Sternus*, *LT224 Justifer*, *LT717 Young Fred* and *LT1172 Constant Star*. Also present is the the Grimsby steam drifter *Treasure*.

The Lowestoft drifter *LT1134 Shipmates* in the South Bay at Scarborough in the 1930s. The well known landmarks of the Grand Hotel and the Spa Bridge can be seen in the distance. *Shipmates* was built by Crabtree & Co. Ltd. at Gt.Yarmouth. Her engine was a 32hp compound made by her builders. She was sunk during a raid by German aircraft on Dover on the 14th. November 1940.

When this photograph was taken on the 13th. October 1934, *YH392 Hilda Cooper* had recently been fitted with wireless. She is seen leaving Lowestoft. Built in 1928 by Cochrane & Sons Ltd., at Selby for Bloomfields Ltd., she was sold in 1956 and was converted to diesel power. *Hilda Cooper* was renamed *Specious* under the ownership of Putford Enterprises Ltd. of Lowestoft. She was sold for scrap to T.G.Darling at Oulton Broad in 1968.

Entering Lowestoft on the 30th. September 1938 is the now preserved *YH89 Lydia Eva.* She is now owned by the Lydia Eva and Mincarlo Charitable Trust Ltd. Clearly noticeable are her original height funnel and mizzen mast. At present these are of reduced height. In 1938 when owned by H. J. Eastick Ltd., she was in her last year as a fishing vessel: soon *Lydia Eva* would be sold out of fishing for ever. She is usually open to the public in the summer months at Lowestoft or Great Yarmouth. The story of the preservation of this vessel has been well documented in a number of books. *Lydia Eva* was built in 1930 by the Kings Lynn Slipway Co. Ltd., as a drifter/trawler. The fitting out was carried out by Crabtree & Co. at Great Yarmouth. Her engine was a 41hp triple expansion steam engine manufactured by Crabtree.

(Top Right) *LT534 Go Ahead.* Built in 1919 by Colby Bros, she was lost on the 18th. November 1940 off Sheerness as *HMD Volume. Go Ahead* is seen here on the 7th. October 1932.

(Bottom Right) *LT767 Heather.* Built in 1917 at Sam.Richards shipyard, she was formerly *YH657*. In this view, taken on 4th. October 1930, *Heather* is making for Lowestoft.

(Above) *LT52 Girl Pamela.* Built in 1912 at Findochy with a F.W. Carver 26hp compound engine, she was formerly *WY169 Oburn* and *YH17*. On the 11th. October 1935 she is seen here leaving her home port. The *Girl Pamela* was lost after a collision off Dunkirk, on the 29th May 1940.

At one time part of the famous Kelsall Bros. and Beechings Ltd. fleet of Hull, with the registration *H108*, the steam trawler *LO249 Grosbeak* had a number of other owners during her life. Having just passed through the Swing Bridge at Lowestoft on the 31st. August 1946, she is about to pass the drifter *LT758 Vera Creina* moored in the bridge channel. *Grosbeak* was built at Goole in 1910 and sold for scrapping in 1954. She was one of a number of similar trawlers working out of Lowestoft at the time. The *Vera Creina* was built in 1911 at Great Yarmouth and sold for scrapping in 1955.

The Fraserburgh motor drifter *FR189 Golden Gain* heads out from Great Yarmouth during the autumn herring season. *Golden Gain* was a regular visitor to Yarmouth in the late 1940s and early 1950s.

86

Owned by Wellbotton (Trawlers) Ltd., *LO496 William Rhodes Moorhouse* leaves Lowestoft in the early 1950s. This former Admiralty vessel was built in 1944 by East Anglian Constructors Ltd. at Oulton Broad as *MFV1557*. Wooden MFVs such as this, played a very important role in the post war development of the fishing industry at Lowestoft. Initially fitted with a 4 cyl. Crossley 240hp diesel engine, she was re-engined in 1961 with a 6cyl. Ruston engine. The *William Rhodes Moorhouse* was sold in 1967 to to Milford Haven. In 1968, she sank in the southern Irish Sea.

A Consolidated Fisheries trawler *LT319 Gunton,* about to leave the Trawl Dock at Lowestoft in the late 1940s. She was built in 1917 at Paisley. *Gunton* was formerly *HMT John Cowarder* and also *GY 289 River Nith.* On the 15th. March 1955, she left Lowestoft for Aberdeen after being sold. There she became registered *A12*. *Gunton* was on the Lowestoft register from 1939 until 1955. In 1960 she was sold for scrap.

(Opposite) Two gents from the era of coal fired trawlers and drifters, at Lowestoft in the late 1930s.

In 1946, Richards Ironworks at Lowestoft launched the 104ft. *Boston Spitfire*, which was ordered by Boston Deep Sea Fisheries Ltd.(BDSF).This was to be the first of a long succession of vessels which Richards would build for that company. By 1962, twenty four vessels had been ordered from Richards by BDSF and its associates. *LT 285 Boston Spitfire* was completed in 1947. She was fitted with a 5 cyl. 330hp. Crossley diesel engine.

In 1953 *Boston Spitfire* was transferred to Acadia Fisheries Ltd., Halifax, Nova Scotia, and renamed *Acadia Fisher*. In 1961, *Acadia Fisher* was wrecked in the Canso Straits. She was declared a total loss (see above)

The wreck of the *Mary Heeley*

The Lowestoft trawler *LT 308 Mary Heeley* was declared a total loss after grounding on rocks in Douglas Bay, Isle of Man. It happened during the night of 29th./30th. April 1950, in dense fog. The *Mary Heeley* was initially the R.N.M.D.S.F vessel *LO197 Edward P. Wills* (see above). She had joined the Lowestoft fleet in August 1949, after being sold by the Mission. The *Mary Heeley* was built at Goole in 1937.

Trawling from Great Yarmouth

In the days of sail, Great Yarmouth was the most important trawling port in the country. In the 1940s and 1950s several large steam trawlers and motor trawlers were owned or managed in Great Yarmouth. These vessels landed at the port and were in addition to the many Great Yarmouth owned drifter/trawlers. Most of the photographs on these two pages were taken at Great Yarmouth or Gorleston.

(Top right) *TA17 Shulasmith* owned by a company in Tel Aviv. She was 170ft. in length and of 216 gross tons. *Shulasmith* was built for fishing the Grand Banks by Hall, Russell in Aberdeen in 1929. During her life she was named *Simon Duhamel, Capitaine Armand,* and later *Kassiopeia.* She was scrapped in 1962 after a spell in Polish ownership. (Bottom left) The Tel Aviv owned, 1933 built *Miriam* was formerly *GY508 Aston Villa* and *FD262 Fotherby.* She also spent some time as the Polish owned *Pollux* and rendered naval service as *Mascot* and later *Vulcan.* She was 140ft. long and of 397 gross tons. (Bottom Right) A rare and remarkable photograph of a large trawler of the Bloomfields fleet. The trawler *A48 Strathalbyn* was 117ft. in length and 218 gross tons. She was owned by the Gt.Yarmouth company for several years, having previously been owned by the Aberdeen Steam Trawling & Fishing Co. Ltd. In 1950 she was sold to Nigeria and during 1955, was reported as being wrecked.

(Below) *GY411 Nacre* was owned by Associated Trawlers Ltd., Gt.Yarmouth. She was built in 1907 at Aberdeen and between 1907 and 1948 was *A191 Strathlin, GN2 , A280 Nacre,* and in 1942 became *GY411. Nacre* was sold for scrap in 1950. (Top right) *GY387 St Clair was* built in 1903 at Hull. She was 128ft. long and 258 gross tons. Previously she was registered *H803* and *FD15.* She was wrecked on the 23rd. August 1949, whilst homeward bound to Great Yarmouth with a large catch caught off the Faroes. *St.Clair* was owned by Associated Trawlers Ltd., Great Yarmouth. (Bottom right) *H571 Avon* owned by Associated Trawlers Ltd., Great Yarmouth. She was built in 1907 at Selby and was 125ft. in length. *Avon* was initially registered *GY340* and of 250 gross tons. She was sold for scrapping in 1950.

Moving down the river at Great Yarmouth late in the afternoon of Saturday 12th. November 1955, is the Fraserburgh drifter *FR87 Xmas Star*. Her principal owner at the time was A. W. Stephen.

The Yarmouth drifter /trawler *YH81 Craiglea* started life as the standard drifter *HMD Rainbrand*. She is seen here leaving Lowestoft whilst in the ownership of Mr.C.V.Eastick. In 1921 *Craiglea* had been registered *LH270* and in 1923 she became *INS540*. *Craiglea* was built in 1920 by the Ouse Shipbuilding Co. Ltd., Goole and sold to a Belgium shipbreaker.

The Banff drifter *BF181 Elm* enters Lowestoft in the 1950s. She was formerly the Admiralty vessel *MFV1225*. In 1959 she was sold and became *BCK118 Crimond*. *Elm* was a regular visitor to Lowestoft for many years.

Although not the last commercial steam trawling vessel to land white fish at Lowestoft, the trawler *LT154 Cairo* was the last commercial steam trawler to operate from the port. In her last few years at Lowestoft she had spells of being laid up at North Quay. The *Cairo* was built at Beverley in 1902 and was initially owned by the Hull Steam Fishing & Ice Co. Ltd. with the registration *H550*. Her nett tonnage was 63 tons and she had a gross tonnage of 172 tons, her length was 108 feet. *Cairo* was built of iron, and for part of her life was owned by the Boston Deep Sea Fishing & Ice Co. Ltd., Fleetwood. Her last owner was Mr. George Mitchell of Lowestoft. On 17th. November 1959 she left her home port for the last time and headed for a breakers yard on the Thames. *Cairo* had made her last landing on the 25th. June 1959.

(Below and Top right) *Cairo* in the Trawl Dock.

(Right) With a question over their futures, the steam trawlers *Cairo* and *Warbler* are seen at the North Quay on the 15th. April 1958. *Cairo* would go to the breakers in 1959, *Warbler* would have a new lease of life as a diesel trawler.

94

The inshore fishing vessels of East Anglia play an important role in maintaining fish supplies. Two representatives of the Lowestoft fleet are shown here. Vessels of the Gt. Yarmouth fleet can be found elsewhere in this book.

(Above) Just returning from trials in 1963, *LT446 G & E* was built by Richards at Lowestoft for G.W. & E.J. Smith. A well known member of the Lowestoft fleet for many years, her dimensions when built were 39ft 6ins. x 12ft. 6ins. x 5ft. 3ins.

(Left) Another well known vessel at Lowestoft seen leaving the port was *LT369 Floreat III,* owned by Terence Catchpole & Robert Pigney. Later to became *Sea Eagle,* she was sold away from Lowestoft to a new owner in St. Just.

(Above) A regular visitor to Lowestoft for many years during the autumn home fishing was *FR287 Valkyrie II*. We see her here crossing the Waveney Dock, in the mid 1950s.

(Left) Two Lowestoft steam drifter/trawlers in the Hamilton Dock in the mid 1950s. On the left is *LT1136 Cyclamen* and on the right *LT129 Jackora*. Just visible is the drifter *LT316 Norbreeze*.
Cyclamen was built in 1911 and sold for scrap in 1957, *Jackora* was built in 1918 and sold for scrap in 1955 and *Norbreeze* was built in 1920 and sold for scrap in 1955. In the background can be seen some of the pickling plots, a hive of activity in the autumn home fishing.

Newly completed and ready to start fishing, the drifter/trawler *LT184 George Spashett* waits for the swing bridge at Lowestoft to open. She was part of the Lowestoft fleet from 1950 until 1965 when she was sold to Longusta Trawling Co., Capetown. Launched on the 15th. August 1950, she started trials on the 2nd. October 1950 and was registered on the 9th. October 1950. Her main engine was a 240bhp 4cyl. Ruston diesel and her dimensions were 81.2ft. x 20.9ft. x 9.0ft. In South Africa she became *Longusta 1*.

(Top left) Many Yarmouth vessels visited Lowestoft to use the dry dock, and slipways for maintenance purposes. In the mid 1950s, the Yarmouth drifter/trawler *YH33 Noontide,* owned by W & L Balls Ltd. is entering Lowestoft for that purpose. *Noontide* was built in 1918 as a standard Admiralty drifter at Oulton Broad. She was previously registered *KY6* and later *KY163*, and was sold for scrapping in 1960.

(Top right) Drifter/trawlers belonging to Bloomfields Ltd., being rigged for trawling at Gt.Yarmouth in the 1950s. Nearest is *YH47 Ocean Dawn,* built in 1919 at Aberdeen.

(Bottom right) The drifter/trawler *YH296 Ocean Hunter* seen leaving Great Yarmouth rigged for trawling, when owned by Bloomfields Ltd. She was built as the Admiralty drifter *HMD Current,* and in 1920 she became *SN42 Current.* During 1922 she was on the move again and became *KY175 Copious.* In 1948 her new owner was Bloomfields and after 4 years as a Yarmouth vessel, she was sold to Edward Beamish and re-registered *LT322. Ocean Hunter* was sold for scrap in January 1955.

Scottish drifters in Great Yarmouth harbour in the mid 1950s. Passing down the river on her way to sea is *FR132 Tea Rose*. Moored outside the cafe is *PD298 Resplendent*.

A superb scene as *YH78 Rosebay* heads for the harbour mouth at Great Yarmouth. This well known drifter was launched on the 13th. November 1919 as the Admiralty drifter *Grey Sea*. She was built at the J. W. Brooke(later Brooke Marine) shipyard at Oulton Broad. Over the years she had numerous English and Scottish registrations before becoming *YH78*. This scene was recorded in 1957. In November 1961, she left Great Yarmouth reportedly bound for a Dutch shipbreakers yard. Her last skipper was Bert Brown, previously he had been mate on her.

A sight still missed today by many in Great Yarmouth. A wooden steam drifter together with the "Eat More Herrings" sign. The drifter which is perfectly positioned with the sign, is the 1927 built *YH92 Achievable,* which until 1931 was *LT 341*. She was laid up in April 1956, when this photograph was taken.

Owned by the Loopey Fishing & Development Co. Ltd., the steam trawler *LT572 Ouse* approaches the harbour entrance at Lowestoft. Earlier in her life she was registered *H514* and owned by Jas. Leyman & Co. Ltd., Hull. Built in Govan in 1900, *Ouse* was sold for scrap in 1954.

Built in 1957 at Gdansk in Poland to a typical Polish design, the 105ft. *YH372 Autumn Star* was one of a similar pair which were registered at Great Yarmouth. The other vessel was *YH370 Autumn Sun,* built in 1956 also at Gdansk. Both of these vessels were classed as lugger/trawlers. The *Autumn Star* was owned by the Herring Industry Board. She was powered by a 360hp 6 cyl. diesel engine. In 1964 *Autumn Star* was sold to Norway and became *HoststjernIa*.

(Above) The drifter fleet leaves port on a Sunday morning in the late 1920s. The drifter nearest is *LT765 Xmas Daisy*; she was scrapped in the early 1930s.

(Right) The scene on a Sunday morning during the home fishing of 1956. The majority of the local drifters have left harbour and three of the last to go, *LT495 Lizzie West, LT207 Feaco* and *YH167 Ocean Sunlight* make their way to the harbour entrance. *Ocean Sunlight* was fishing from Lowestoft as a trial instead of her home port.
The Scottish drifter in the background would sail next day with the rest of the Scottish fleet which is out of sight in the Hamilton Dock. The Scottish drifter fleet stayed in port on Sundays.

Many former Admiralty MFVs could be seen at Lowestoft and Gt. Yarmouth in the 1940s, 1950s and 1960s. In addition to the local drifters and trawlers, many visiting Scottish drifters were former MFVs. Viewed from the now closed ferry in August 1956, is *YH384 Scadaun*, one of these vessels. She was built in 1945 at Peterhead. At the time that this photograph was taken, *Scadaun* was laid up. Later she was sold to Aberdeen and became *A103 Greenbank*.

A fine scene at Gorleston as *YH92 Achievable* sets out for sea. Built at the Oulton Broad yard of John Chambers in 1927, she was initially *LT341*. Her engine was a 34hp triple expansion built by Elliott & Garrood, and her dimensions were 87ft. 7ins. x 19ft. 10ins. x 9ft. 9ins. She had a gross tonnage of 96 tons. *Achievable* was sold for scrap in 1957.

The Lowestoft drifter/trawler *LT337 Ethel Mary* rigged for trawling, leaves Lowestoft for the fishing grounds. Built by Richards at Lowestoft in 1957, she was sold in July 1969 to A. J. & A. Buchan of Fraserburgh. As the *Golden Promise* she embarked on a trip to the South Atlantic, Capetown and Tristan da Cunha. This was reportedly to fish for crayfish. She was later sold and became *PD250 Magnificent.*

(Above) Visiting drifters in the Hamilton Dock at Lowestoft. The Lerwick drifter *LK511 John West* and the Penzance drifter *PZ85 Karenza* can seen.

(Top right) A typical Sunday morning during the home fishing in the 1950s at Lowestoft. It was usual for people to line the piers and other vantage points to watch the local drifters head out for the herring grounds. *LT126 Loyal Friend* was built locally at the yard of J. W. Brooke as *HMD Low Tide*. She was launched on the 3rd. December 1919. Other names which *Loyal Friend* had during her life were *Seaward* and *Mary Flett*. She left her home port for a breakers yard in Belgium on the 17th. May 1957. Her last local owner was J. J. Colby. In the background is the drifter *LT224 Justifier*.

> Prices on Saturday reached 158s. a cran for fresh. The top landing was that of the Banff drifter Daystar which landed 38 crans. Other landings included 30 crans each by Henrietta Spashett (LT), One Accord (LT), Prevail (BF), Silver Wave (BF) and Golden West (BF). On Monday the price went to 250s. a cran when only 14 boats returned with just under 160 crans. Norfolk Yeoman, of Lowestoft, was top with 34. Loyal Friend had 28. Young Duke 18. Harold Cartwright 16 and Swiftwing 14.

(Left) A 1956 landing report. Reproduced here by kind permission of Mr. Tim Oliver, Editor of the Fishing News. From the Fishing News of the 9th. November 1956.

LT365 Merbreeze earned a place in local maritime history in 1931, by being the last steam drifter to be built by Sam Richards shipyard. She was also the last coal burning drifter/trawler built for Lowestoft. This 93ft. drifter/trawler was built for P.W.Watson & Sons Ltd. Initially registered *LT253, Merbreeze* was converted to diesel in 1959, and sold for scrap in 1976. This photograph was taken on the 14th. October 1956 as she was leaving Lowestoft.

Seen leaving Lowestoft is the Yarmouth drifter *YH225 Fortitude,* owned by Mr. Ronald Balls. This 72ft. vessel was previously Admiralty *MFV1242*. She was built in 1945 at Grimsby, by J.S.Doig. In 1961 she was sold to a company which intended to start a fishing fleet in Morocco. *Fortitude* was at Lowestoft in March that year, being prepared for the long voyage to the Mediterranean. At the time Mr. Balls said the sale was " a sign of the times. The small owner can't afford to keep going these days ". With just one drifter he had found it impossible to get a crew together, most people preferring to work for a large company.

The trawler *H587 Warbler* was built by Goole Shipbuilding & Repair Co. Ltd. and completed in 1912. She was built for Kelsall Bros. & Beeching Ltd. of Hull, and was 71 tons net and 192 tons gross. Her dimensions were 110ft. x 21.65ft x 12.05ft. In 1940, *Warbler* was sold to Brandon Fishing Co. Ltd. and in 1958 passed to W. H. Podd. Ltd. Conversion to diesel took place in 1958/59 with the installation of a 500hp 5 cyl. A. K. Diesel engine. The work was carried out by LBS Engineering. In 1968 *Warbler* was sold to P.F., F.E., and G.A. Catchpole and J. R. Hashim. She was converted for safety standby work. In 1972 *Warbler* was sold to T. G. Darling Ltd. for scrapping.

(Above) Looking across the deck of the Yarmouth drifter/trawler *Ocean Trust,* on the 7th. February 1959, we see the *Warbler* after conversion. She is in the Inner Harbour at Lowestoft, and has her new registration of *LT63*.

(Right) Having just arrived back from a fishing trip, the *Warbler* lies gently hissing and sizzling in the Trawl Dock at Lowestoft, on the 21st. March 1957. In this view she has her London registration *LO251*, and painted on her funnel, the Short Blue flag. This was the house flag of the famous Hewett group of fishing companies. The *Warbler* is best remembered as seen here.

In 1958, the Yarmouth registered drifter/trawler *YH84 Deelux* owned by W.H.Kerr (Ship Chandlers) Ltd., was converted from steam power to diesel. The conversion was carried out by Richards Ironworks Ltd. at Lowestoft. (Bottom left) Early in 1958 work has just started. (Top Left) The 30th. November 1958, and the conversion has been completed. She is in the Trawl Dock at Lowestoft. (Main Photograph) A few years after the conversion, and she is earning her keep as a diesel trawler, heading out for the fishing grounds. *Deelux* was built in 1930 at Aberdeen by Alexander Hall & Co. Ltd., for Bloomfields Ltd. of Great Yarmouth as the *Ocean Lux*. During the conversion a 360hp 5 cyl. Crossley diesel engine replaced the 39hp 3cyl. A. Hall triple expansion steam engine she was originally powered by. In 1975, *Deelux* was sold to T.W.Ward Ltd. for scrapping.

...sel was built for the Admiralty as *MFV 1534*. During 1947 she was ...y a Grimsby fishing company and became *GY510 Grasby*. 1950 saw ... to the Huxley Fishing Co. Ltd., Lowestoft, and re-registered *LT267*. ... the Lowestoft fleet in 1955, after being sold to Aberdeen where she ... *A66 Doonie Braes*. *Grasby* was built at Charlestown in 1947.

The 16th. January 1961, and the drifter/trawler *LT245 J.A.P.* is aground in the Inner Harbour at Lowestoft. She was built in 1931 at Richards shipyard at Lowestoft. A twin screw vessel, *J.A.P.* sank on the 19th. February 1967, near *Smiths Knoll* L.V. In the background, moored at the Sleeper Depot is *D411 Loch Lein* owned by Claridge Trawlers Ltd., she sank in 1965 off the coast of Australia.

110

(Opposite top left)
LT345 Primevere owned by J.J. Colby leaving Lowestoft on a typical Sunday morning in the herring fishing of 1957. Built in 1914, she was sold for scrap to a Belgium shipbreaker in 1960.

(Opposite bottom left)
LT7 Shepherd Lad leaving Lowestoft on a Sunday morning in 1957. On the 6th. Dec. 1960 she sailed bound for a shipbreaker in Belgium.

(Opposite right)
A study of the *LT167 Hosanna* taken on the 27th. December 1959, in the Hamilton Dock, Lowestoft. In 1938 she was the winner of the Prunier Trophy. *Hosanna* is waiting for work to start on her conversion from being steam powered to diesel.

(This page)
The result of the conversion mentioned above. *Hosanna* is in the Waveney Dock, Lowestoft. She was sold for scrapping in 1976.

Crossing the Waveney Dock on the 2nd. April 1956 is the trawler *LT1166 Eager,* built by Cochane & Sons Ltd. at Selby as a steam drifter/trawler in 1912. She was converted to diesel power in 1953 by LBS Engineering Co. Ltd. Whilst at Lowestoft, *Eager* passed through the ownership of S. Allerton, F. Spashett, Bay Fisheries, H. Roberts, Eager Fishing Co., W. H. Podd and Diesel Trawlers Ltd. She sailed for Ghana in 1963 after being sold. *Eager* was reported in 1972 as having sunk.

The Waveney Dock at Lowestoft has seen many changes, including the types of vessels using it and the use made of it. However, the demolition of the buildings erected by the Great Eastern Railway was the most drastic of the changes. Some of the buildings are visible in these views.

The morning of Sunday 9th November 1958, with drifters preparing to leave port. The majority are those of the Small & Co. (Lowestoft) Ltd. fleet. Identifiable are *LT231 Harold Cartwright, LT82 Henrietta Spashett, LT137 Norfolk Yeoman.* Others are *LT46 Silver Crest* and *LT365 Merbreeze.*

(Bottom right) The morning of Sunday 29th. October 1961, and *KY322 Wilson Line,* owned by C.V.Eastick leaves Lowestoft. *Wilson Line* was converted in 1959 from steam to diesel power. In 1962 she was re-registered *YH105,* after the previous vessel with that registration, the *Wydale,* was sold for scrap. *Wilson Line* was built in 1932 at Aberdeen, she was sold to Greece in 1975.

(Top left) Leaving Lowestoft on the 21st. January 1938, is the Ramsgate registered drifter/trawler *R129 Mill O'Buckie*. Not a common sight in Lowestoft, both this vessel and *R355 Lady Luck* were in the harbour on the 3rd. March 1957. Their owners had become part of the Colne group. Both vessels left during that month after being sold for scrap.

(Top right) Until 1963, *LT465 Ocean Starlight* was *YH61*. She was sold to Holland in 1967. On return to Lowestoft in 1972 she became *Stoic* and in 1981 the *Dawn Spray*. She was used on safety standby work until 1987 when she was sold to Milford. In 1962 as *YH61 Ocean Starlight* she was winner of the Prunier Trophy. On top of her mizzen mast can be seen the weathervane presented to the vessel on that occasion. In 1995 *Dawn Spray* was sold for scrap.

The 4th March 1961 and *HL48 John O'Heugh* leaves Lowestoft. Powered by a 400hp 6cyl. Crossley diesel engine, *John O'Heugh* was bought in 1963 by Pegasus Trawling Co. Ltd, and became *Boston Trident*. In 1972 she was sold to Safety Ships Ltd, becoming *Carbisdale*. In 1979 *Carbisdale* was sold for scrapping to a Blyth shipbreaker.

The morning of Sunday 29th October 1961, and the drifter/trawler *LT177 Kindred Star* leaves for the fishing grounds. *Kindred Star* was converted from being steam powered to diesel in 1954 by LBS Engineering Ltd. She was built at Oulton Broad in 1930 and was owned by the Star Drift Fishing Co. Ltd. *Kindred Star* was sold in 1965 and left the port in 1966.

Prominent amongst the fishing fleet of Lowestoft for over fifty years were the large number of vessels owned, or at one time owned, by the Lowestoft Steam Herring Drifters Co. Ltd. The names of most of their vessels had the prefix "Lord". Four vessels which were units of that fleet are shown here, the final known end for these vessels was in each case different.

(Right) *LT20 Lord Hood* and *LT55 Lord Barham* await their fate on the 2nd. November 1957, moored at the Lowestoft Sleeper Depot (Now the Shell base). Both were built at Selby in 1925. *Lord Barham* was sold for scrap in 1960. She was formerly *LT211* and *GDY110 Arkadiuz*. *Lord Hood* was sold for use as a salvage vessel and left for Aberdeen on the 27th. April 1959. She was previously *LT215* and *GDY108 Antonieuz*. In 1960 she was sold for scrap. Her claim to fame was in 1952, when she won the Prunier Trophy under Skipper Ernest Thompson. (Bottom left) *LT79 Lord St.Vincent* was built by John Chambers Ltd. in 1929. She was mined and sunk in the Thames Estuary on the 7th July 1941. This scene was recorded on the 28th. October 1936. (Bottom right) The drifter/trawler *LT1141 Lord Haldane* was built by Cochrane & Son in 1911, she was 84ft. in length and had a gross tonnage of 91 tons. In 1940 she was declared a total loss after going missing in the Bristol Channel area. In this very early photograph she is seen hauling on the herring grounds.

This fine view of the trawler *LT457 Rosevear* is from the archives of Brooke Marine Ltd. Built in 1962 at their Oulton Broad shipyard as yard no. SYC282, she was one of four similar vessels built at the yard.. Three of these were built for Lowestoft, the other for Aberdeen. One of her sisterships is the now preserved *LT412 Mincarlo,* featured elsewhere in this book. *Rosevear* had a number of owners, and worked out of Milford Haven for fourteen years. She was sold to a shipbreaker in 1985.

LT203 Annrobin was built by Richards Ironworks as yard no. 426, for the East Anglian Ice & Cold Storage Co.Ltd. in 1955. In this photograph from her builders archives she is off the harbour entrance at Lowestoft. In 1968 *Annrobin* was sold to Italy and became *Mary*.

Formerly registered *GY513* and later *M204 Milford Baron, LT134 British Honduras* was built at Selby by Cochrane & Son in 1937. She was fitted with a Ruston diesel engine, and owned at Lowestoft by Claridge Trawlers Ltd. Sold for scrapping in 1968, she left the port in August that year with *LT186 Kingfish* in tow. They were bound for a shipbreakers yard near Sheerness.

The approach to the harbour entrance at Lowestoft, with a trawler and Scottish drifter leaving the port. The trawler is *LT186 Kingfish,* built in 1955 at Thorne and the first of the early batch of "Fish" class trawlers. These were built for subsidiary companies of the Colne group. The overall length of *Kingfish* was 102ft. 10ins. She was fitted with a Ruston 6VEBM engine giving 338shp at 550rpm. *Kingfish* was used on offshore safety standby work for a number of years, before being sold for scrapping in 1986.

The day after the Christmas break and the Lowestoft trawler fleet leaves port, this scene was recorded in the mid 1960s when the fleet numbered over 120 vessels. *LT459 Boston Coronet* was previously registered *GY596*. She was powered by a 550hp 5cyl. Widdop diesel engine and had a gross tonnage of 199 tons. In 1980 *Boston Coronet* was sold and became *Lidia Prima*. Also identifiable in the photograph is *LT155 Filby Queen,* and on the right the stern trawler *A344 Universal Star,* the first of its type at Lowestoft.

The Lowestoft trawler *LT432 Boston Pionair* was lost with all hands in 1965. The last report from the vessel was received around 0630hrs. on the 14th. February 1965. In the days following a number of items of wreckage were found from her, these included a liferaft and lifebuoys. She is seen here entering her home port with the registration given to her in 1962. Originally her registration was *FD96*. The *Boston Pionair* was owned by Pegasus Trawling Co. Ltd., a Boston Deep Sea Fisheries Ltd. subsidiary company. She was built in 1956 as yard no. 429 by Richards Ironworks Ltd., Lowestoft. Her main engine was a 6 cylinder 500hp Widdop diesel, she was one of many trawlers built to a similar design by Richards.

Leaving the Trawl Dock at Lowestoft on the 6th. May 1961 for sea is the trawler *LT67 Anguilla*. When taken off fishing, she spent sometime as a safety standby vessel. During 1986 she was sold to Milford Haven. *Anguilla* was sold again in 1992, and by 1997 had been converted into a sailing vessel. When built in 1959 at Selby, *Anguilla* was 104ft. in length and 228 gross tons. She was built for the Clan Steam Fishing Co.(Grimsby) Ltd., a Colne group subsidiary. Later she was transferred to Claridge Trawlers Ltd., another subsidiary within the same group.

The former Admiralty *MFV1536* and later Grimsby trawler *GY531 Harrowby,* became *LT317 Vesper Star* in the early 1950s. She is seen here as a drifter leaving Lowestoft on the 12th. October 1958, on her way to the fishing grounds. *Vesper Star* was owned by the Star Drift Fishing Co. Ltd. In 1961 she was sold and re-registered at Shoreham.

The L.F.V.O.A. Landing Order Board, situated near the entrance to the Waveney Dock, was a feature of Lowestoft for many years.

On this Monday in 1967, twelve trawlers landed, including *LT59* (right). Reference to the book " Down The Harbour 1955-1995" will reveal the identity and details of all the vessels listed on the board

Yard No. 422 at Richards Ironworks Ltd. in 1954 was *LT59 Diadem,* ordered by W.H.Kerr (Ship Chandlers) Ltd. In 1963, *Diadem* became *Boston Caravelle* after being transferred to Boston Deep Sea Fisheries Ltd. Ownership was transferred a number of times between different companies within Boston Deep Sea Fisheries Ltd. Eventually in 1973, she was sold and became part of the fleet of Safetyships Ltd., Aberdeen. *Boston Caravelle* was renamed *Dunnichen*. During 1979 she was sold to H. Kitson Vickers & Sons Ltd., Blyth, for scrapping. *Diadem* had dimensions of 103ft. x 22ft 1ins. x 10ft. 9ins. She had a gross tonnage of 166 tons and was fitted with a 6 cyl. 440hp Crossley diesel engine when built.

The Eastick family were very well known fishing vessel owners for at least five generations. Featured here are just two of their vessels, others will be found elsewhere in this book.

YH278 Harry Eastick was built in 1926 at Great Yarmouth and sold for scrapping in April 1961. She is seen here on the 27th. September 1958 at Lowestoft.

The last steam drifter at Lowestoft, *LT495 Lizzie West* heads out of Lowestoft on the 28th October 1958. Built at Buckie by Herd & McKenzie in 1930, her last local owner was C.V. Eastick Ltd. of Gorleston. She was sold for use as a tanning vessel in 1961. By 1968 she was considered life expired and beached. *Lizzie West* was initially registered *BF213*, her principal owner being John West of Gardenstown. She was sold after the Second World War and re-registered *M22*, before finally becoming *LT495*. For many years *Lizzie West* was owned by members of the Mitchell family.

A common sight in Lowestoft are the research vessels operated by the Ministry of Agriculture, Fisheries and Food. Two of their well known vessels, which were very active in the 1940s and 1950s, are featured here.

(Right) The disposal in 1961, of the well known research vessel *LT263 Sir Lancelot*.

Reproduced by kind permission of Tim Oliver, Editor of Fishing News.

OFFICIAL NOTICES
ADMIRALTY SMALL CRAFT DISPOSALS
The Director of Navy Contracts is prepared to consider offers for the following craft. The closing date for receiving offers is noon Tuesday, 25th April, 1961, and tender forms and Conditions of Sale Form S.C.D.117a can be had on application to Director of Navy Contracts, Branch 8D (1), Admiralty Offices, Ensleigh, Bath. Telephone number Bath 6933 Ext. 1184.
RESEARCH TRAWLER VESSEL "SIR LANCELOT"
Lying at Donaldson Quay, Commercial Road, Lowestoft. Permission to view by prior arrangement only with Capt. H. J. Aldiss, Fisheries Laboratory, Lowestoft. Telephone Number PAKEFIELD 251.
Steel construction. Built 1942 by John Lewis of Aberdeen. Length 126.2ft. Breadth 23.65ft. Depth 12.8ft. Gross tonnage 295.77. N.R.T. 88.5 Triple expansion steam engine (aux. Ruston & Hornsby), B.H.P 600. One oil fired boiler 200 lbs. W.P. Two 110 D.C. Volt Dynamos (one 15kW and one 5 kW) Cruising speed 9.5 knots. Fuel oil tank: 60 ton capacity. Classified Lloyd's 100% A.1.
Port letter and number : LT/263.
Official No. 166703.

(Above) After arriving at Lowestoft on the 25th. March 1946, the *HMT Sir Lancelot* was converted to become the fishery research vessel *LT263 Sir Lancelot*. She was built at Aberdeen by John Lewis & Co. After many years being based at Lowestoft, *Sir Lancelot* was sold and left the port in February 1962, bound for Germany. She became the *Hair Ed Din Barbarossa*. This scene on the 11th. September 1960 shows her unusually in the Trawl Dock. *Sir Lancelot* was the first of the famous " Round Table" trawler minesweepers to be completed in March 1942.

(Left) This photograph from the archives of her builders, shows *LT205 Platessa* on trials off Lowestoft soon after completion. Built as Admiralty *MFV1576*, she was converted to a fishery research vessel in 1946. Sold out of research in 1968, and eventually hulked at Lake Lothing, her remains are still to be seen there today.

Richards Ironworks at Lowestoft built the great majority of the last drifters/trawlers to be built. Between 1949 and 1960 they built eighteen of these vessels for various owners. Two of these vessels are shown here. Both vessels were fitted with Ruston engines, and were winners of the famous Prunier Trophy.

(Above) *YH61 Ocean Starlight* was built in 1952 and became *LT465* in 1963. In 1967 she was sold to Holland, later to return and work as a safety standby vessel. She was sold for scrap in 1995. This photograph is from her builders archives. (Left) *LT61 Dick Whittington* was built in 1955. She left Lowestoft on the 14th June 1968 after being sold to Quincy & Asaro, Trapani. *Dick Whittington* was renamed *Saturno II*.

The following three photographs were taken on board one of the six side fishing trawlers built at Appledore in the late 1960s, for Small & Co.(Lowestoft) Ltd., and the East Anglia Ice and Cold Storage Co. Ltd. *LT555 Suffolk Challenger* was one of these vessels, and is here about to enter the Waveney Dock at Lowestoft. She was converted to a safety standby vessel in 1981. Just before Christmas 1986, she sailed from Lowestoft for her new role under Anglo Spanish ownership. *Suffolk Challenger* later became *Jer Dos*.

Looking out of the wheelhouse, with the vessel trawling. These highly efficient top earning trawlers, were of 255 gross tons and 125ft. overall. They were built with the emphasis on safety and crew comfort. All six of the class were sold to Anglo Spanish interests in 1986/7.

Two further views taken from the wheelhouse.

(Right) At Great Yarmouth, the afternoon of Saturday 12th. November 1955 was dull, misty and with a little drizzle. *YH78 Rosebay* along with other drifters is at rest, waiting for the following morning when she will once again head out from her home port in search of herring.

(Bottom left) Returning from a fishing trip, the drifter/trawler *LT375 Young Elizabeth* ready to unload, enters the Waveney Dock at Lowestoft. She was built in 1953 at Richards Ironworks in the town. In 1968 she was sold out of fishing and initially used as a diving support vessel.

(Bottom right) Four Scottish drifters make for the harbour entrance at Lowestoft, led by the Peterhead drifter *PD267 Fertility,* a regular visitor to Lowestoft for many years. She was one of the few Peterhead drifters to work from the port.

Over the years large numbers of steam and diesel trawlers from the Humber ports have moved to Lowestoft. The 120ft. *LT372 Suffolk Craftsman,* seen here leaving for the fishing grounds, was previously *GY672 Priscillian,* owned by Dominian Steam Fishing Co. Ltd. She was bought by Small & Co.(Lowestoft) Ltd. in early 1977. Built at Selby in 1961, *Suffolk Craftsman* left Lowestoft in 1980 after being sold to Greece where she became *Ion.* She was reported as being scrapped in 1984. This vessel was the second *Suffolk Craftsman* to be part of the Small & Co. fleet. The first was sold in 1974, later to become *Winkleigh.* Close examination of this scene will reveal the Landing Order Board featured elsewhere in this book.

The Grimsby trawler *GY616 Saxon Venture* became *LT165 Tobago* when she was bought by the Colne Fishing Co. Ltd. She is recorded here leaving Lowestoft, and in the background can be seen *LT83 St.Nicola* and *LT32 Bentley Queen*. As built, *Tobago* was 103ft. long and had a gross tonnage of 211 tons. In 1980 she was converted for safety standby work, and in early 1987 was sold for scrapping at New Holland. She was the second Colne group vessel to carry the name *Tobago*. The first was towed away to the shipbreakers in 1964 after spending sometime aground, on the Old North Extension at Lowestoft.

SN36 Blacktail was sold in 1965 by Pelagic (Realisations) Ltd., to Talisman Trawlers (North Sea) Ltd. and became *LT 502 Farnham Queen*. She is seen here entering Lowestoft harbour. Built by T. Mitchison Ltd. at Gateshead in 1961, *Farnham Queen* was sold in May 1983 to Anglo-Spanish interests and became *LT502 Summer Swallow*. She was sold again in 1993 and became *M425 Sea Dog*. The *Sea Dog* sank on the 16th. December 1998 approximately 200 miles west of Ireland, the crew took to the liferafts and were rescued. The original main twin 4cyl. Maybach engines were replaced during 1967 with twin 8cyl. Paxman engines. These were replaced by one 6cyl. 850hp. English Electric engine in late 1969. As built, *Blacktail* was 114ft. in length and of 246 gross tons.

An early colour photograph taken at Great Yarmouth showing a wooden steam drifter belonging to the Eastick family, with another wooden steam drifter unloading alongside.

An impression by the Great Yarmouth artist K.E.Hemp of the Yarmouth drifter/trawler *YH84 Ocean Lux*. This painting illustrates the very much admired colour scheme adopted by her owners, Bloomfields Ltd., for their vessels. *Ocean Lux* is featured elsewhere in this book as the diesel trawler *Deelux*.

Drifters moored at Gt. Yarmouth on a Saturday afternoon during the autumn fishing in the late 1950s. The drifters are *YH278 Harry Eastick* (nearest) and two of the Bloomfields fleet.

Great Yarmouth and Lowestoft drifters at Aberdeen in the 1950s. Six East Anglian drifters can be identified, two of which are from the Bloomfields Ltd. fleet of Great Yarmouth. The others are *LT152 Thrifty*, *LT137 Norfolk Yeoman*, *LT246 Fellowship* and *LT156 St.Luke*.

A scene from 1958 with drifters in the Waveney Dock at Lowestoft. The vessels present are *LT207 Feaco,* which had been converted to diesel power in 1955, and was sold to Ghana in 1967, *LT365 Merbreeze* which was converted to diesel power in 1959, and sold for scrapping in 1976, *LT77 Prime* the last steam steel fishing vessel at Lowestoft, sold for scrap in March 1961, and *LT345 Primevere* sold for scrap in 1960. In the foreground is one of the three drifter/trawlers which were former Admiralty MFVs, and at the time were herring fishing. Also visible is a coal lighter servicing another steam drifter.

(Above) The last operational steam drifter was *YH105 Wydale*. This scene was recorded shortly before she made her final voyage to a breakers yard in Holland in October 1961. She was built by John Chambers Ltd. at Oulton Broad in 1917. Many say she should have been saved for preservation; for over fifty years this class of wooden vessel played a very important part in the fishing industry of the United Kingdom. *Wydale* won the Prunier Trophy in 1950. In this view the weathervane awarded at the time is clearly seen at the top of the mast.

(Top and Bottom right) Great Yarmouth has been a major British holiday resort for many years. It has always been popular with visitors to Great Yarmouth, to send postcards such as these examples, to relatives and friends. The 1950s scene in the top photograph, shows the Peterhead steam drifter *PD338 Litchen* nearest the camera. The local drifter *YH293 Ocean Unity* is further along. She was owned by Bloomfields Ltd., and was built as a standard drifter in 1920 at Hook, Goole.

136

Seen here at Lowestoft in 1967, *LT310 W.F.P.* was built in 1957 at the Fraserburgh yard of Thomas Summers Ltd. One of a pair built at the yard, she was the first new wooden trawler to join the Lowestoft fleet for many years. *W.F.P.* was built for Lowestoft Motor Trawlers Ltd., part of the W.H.Podd group. According to the builders specification she was 97ft. x 22ft. x 10ft 6ins., with all machinery, including the 3 cyl. 300hp engine supplied by A. K. Diesels Ltd. *W.F.P.* was repossessed by the White Fish Authority in 1968, and sold to Putford Enterprises Ltd. in 1969. She left the port in 1970 and eventually became a total loss during the early 1970s in the Mediterranean. Just visible on the adjacent slipway is the Gt.Yarmouth pleasure vessel *Eastern Princess.*

LT88 Ormesby Queen was built in 1954 at John Lewis & Sons. Ltd. shipyard at Aberdeen. She was fitted with a 6cyl. 440hp Crossley diesel engine. During 1970 *Ormesby Queen* left her home port for a breakers yard at Blyth.

Mr. J. J. Colby of Lowestoft owned or partially owned many fishing vessels over the years. Two of his drifters are featured here. Both are remembered for very special reasons.

The last operational steel built steam drifter was *LT77 Prime*, seen here entering Lowestoft harbour in the last years of her life. Until 1944, *Prime* was registered *BCK185*, between 1944 and 1954 she was *A572*, and she then became *LT77*. *Prime* made her last voyage on the 8th. March 1961 when she left Lowestoft for a shipbreakers yard in Belgium. She was built by the Torry Shipbuilding Co. Ltd. in Aberdeen in 1914. Her dimensions were 87.9ft x 19.6ft. x 9.0ft., and she had a gross tonnage of 101 tons. *Prime* was powered by 52hp(260ihp) triple expansion engine.

LT382 Wisemans is remembered as the very last English drifter to work from Lowestoft and Great Yarmouth. This scene was recorded on the 14th. October 1968 as she left Lowestoft for what had traditionally been the herring grounds. Unfortunately there were few herring there to be caught. By the end of that month she would end herring fishing for ever and be put up for sale. The 1968 autumn herring fishing was a very sad occasion with very few Scottish drifters taking part. Those that did had left for home by the end of November. The total catch of under 1000 crans was the smallest on record. *Wisemans* was sold to Waterford, Eire and was re-registered *W38*. She sank on the 11th. March 1984 in the Irish Sea. *Wisemans* was formerly the Banff drifter *BF154*, and prior to that, the Admiralty vessel *MFV 1025*. She was built by George Forbes at Peterhead in 1943.

The large side trawler *M3 Milford Duke,* was built by Cochrane & Son Ltd. at Selby in 1949 for the Milford Steam Trawling Co. Ltd. In 1955 she was sold to France and became *Jean Vauquelin* . She was bought by Claridge Trawlers Ltd. in 1968 and became *LT82 St.Rose.* In May 1985 her engine was removed and she became the *Unda.* As such she was towed to the shipbreakers yard together with *LT312 Barbados* in July 1985. *St. Rose* is seen here on the 14th. October 1968 off Lowestoft.

Lowestoft Trawl Market in March 1965.

LT397 Suffolk Kinsman, was launched at Richards Ironworks shipyard at Lowestoft on the 11th. July 1960. Her main engine was a 550hp 6 cyl. Ruston, she was 116ft. in length and of 202 gross tons. In 1974, Small & Co.(Lowestoft) Ltd. sold her to Boston Deep Sea Fisheries Ltd. In their ownership she became *Boston Kinsman*. She was sold by them in 1978 and became *Nuovo Diodoro*.

Another product of Richards Ironworks, *LT445 Boston Beaver* was built in 1962 for Boston Deep Sea Fisheries Ltd. She was powered by a 5 cyl. 475hp National engine. In 1978 she was sold to Breydon Marine Ltd. and became *Breydon Mallard*. Her fishing registration was cancelled and she was converted for safety standby work. During 1987 *Breydon Mallard* was sold and re-registered for fishing.

Drifter/trawlers were always prominent at both ports. Diesel powered vessels of this versatile type are featured here.

The Yarmouth drifter/trawler *YH377 Ocean Trust,* rigged for trawling, leaves port in the 1950s. She was built by Richards at Lowestoft for Bloomfields Ltd., and was the last of the six drifter/trawlers built for Gt. Yarmouth. In 1963 she was sold to Small & Co.(Lowestoft) Ltd. and became *LT469. Ocean Trust* spent a few years under Scottish ownership and then returned south to work on safety standby duties. In July 1987 as the *Celita,* she passed into Anglo-Spanish ownership.

The last owner of the former steam powered drifter/trawler *LT188 Tritonia* was Mitchells Tritonia Ltd. Built in 1930 at Oulton Broad she also ended her days there, in 1976 at a shipbreakers yard. The diesel powered *Tritonia* is seen here leaving her home port in October 1968.

On page 106, the drifter/trawler *LT365 Merbreeze* is seen prior to conversion to diesel power. In this view, *Merbreeze* is entering Lowestoft harbour as a diesel trawler on the 4th. June 1974. She was converted from steam to diesel power in 1959. Her owner prior to being sold for scrap in 1976 was Merbreeze Ltd., one of the Mitchell Bros. companies.

Wilson Line is perhaps best remembered having the registration *KY322*. She was re-registered *YH105* in 1962. Built in 1932 at Aberdeen, *Wilson Line* was converted from steam to diesel power in 1959. In 1975 she sailed from Lowestoft bound for Greece after being sold. Her last local owner was Breydon Marine Ltd., Burgh Castle. This view shows her owned by C.V.Eastick.

LT83 St.Nicola was built by Cochrane & Son at Selby in 1949 for the Milford Steam Trawling Co. Ltd., as *M16 Milford Duchess*. During 1954 she was sold to France and became *Joli Fructidor*. The vessel was bought by Colne Fishing Co. Ltd. in 1968 and became *LT83 St.Nicola*. In early 1985 she was renamed *Willem Adriana* and as such was towed away for scrapping in May 1985.

Built for Jackora Ltd. in 1962 by Richards Ironworks Ltd. at Lowestoft, the 93ft. *LT434 Jacklyn* was sold in 1975 to the Colne Fishing Co. Ltd. She was renamed *Barbuda*. Eventually used on safety standby work, *Barbuda* was sold for scrap in 1991. As *Jacklyn,* she is seen here on the 25th. July 1973. Jackora Ltd. was one of the Mitchell Bros. companies.

Another vessel owned by Jackora Ltd. was *LT341 Oceanbreeze*. Converted from steam to diesel power in 1958, she was sold to Greece in 1975. *Oceanbreeze* was built by John Chambers Ltd at Oulton Broad in 1927. She is here entering Lowestoft on the 2nd. April 1973.

The drifter/trawler *LT246 Neves* was converted in 1955 to diesel power and also renamed *Fellowship*. Owned for the majority of her life by the County Fishing Co. Ltd., her last owner was J.J.Colby. *Fellowship* is seen here in October 1968 off Lowestoft. After a period as a safety standby vessel, she was sold and left Lowestoft in 1974. *Neves* was built in 1931 at Goole.

Crossing the Waveney Dock at Lowestoft on the 25th April 1973 is the 1960 built *LT364 Boston Argosy*. A product of Richards Ironworks, she was powered by a 6cyl. 515hp National engine. The ownership of her was transferred a number of times between various companies of the Boston group. *Boston Argosy* was converted for safety standby work in 1972, and reported as being sold to Anglo-Spanish interests in July 1986.

The 1959 built *LT64 Montserrat,* seen here on the 2nd. April 1973, was owned by Huxley Fishing Co. Ltd., part of the Colne group. Like a great many vessels of that group, she was built at Selby by Cochrane & Son. Powered by a 6cyl. 403hp Ruston engine, she had dimensions of 104ft. 7ins x 23ft. 7ins. x 8ft. 4ins. During 1976 she undertook safety standby work, and in 1987 *Montserrat* left Lowestoft after being sold.

As the years pass this vessel is destined to become increasingly well known as an example of a side fishing trawler. The now preserved *LT412 Mincarlo* is owned by the Lydia Eva & Mincarlo Charitable Trust Ltd. The full story of this vessel can be found in the book " Down The Harbour 1955-1995 ". In this view, she is leaving Lowestoft on the 23rd. July 1973 for the fishing grounds, when owned by Putford Enterprises Ltd. At present, work is required on the *Mincarlo* to fully restore her to the condition seen here. This includes reinstating the small boat, one mast, two trawl gallows and removal of major items installed when she was a safety standby vessel.

Seen off Lowestoft on the 15th. October 1968 is the Small & Co. (Lowestoft) Ltd., trawler *LT378 Suffolk Mariner*. Sold in 1975 to Pentode Ltd., an associate company of Boston Deep Sea Fisheries Ltd, her name became *Boston Mariner*. She was sold to Durban in 1980 and became the *Mariner*. *Suffolk Mariner* was built at Lowestoft by Richards Ironworks in 1961.

Built as *FD176 Princess Royal* by John Lewis Ltd. at Aberdeen in 1952, the *Boston Lancaster* came to Lowestoft as the *Acadia Heron*. She had been working off the Canadian east coast for Acadia Fisheries Ltd. A vessel which had many owners and name changes during her fishing career, she was sold in 1970 to a company involved in offshore work. *Boston Lancaster* was renamed *Corsair*. She was sold on in 1973, renamed *Subsea Corsair* and used on diving support work. When built her dimensions were 124ft.7ins. x 25ft.2ins. x 12ft.5ins. Her main engine was a 6cyl. 660hp Mirrlees.

Seen here in 1967, the drifter/trawler *LT671 Suffolk Warrior* sank on the 15th. February 1969, approximately 130 miles from Cromer. She had been in collision with the Dutch Trawler *KW81 Hendrike Johanna*. Built in 1960 by Richards, she was the last vessel of her type to be built. *Suffolk Warrior* was owned by Small & Co. (Lowestoft) Ltd., and was 92 ft. in length and had a gross tonnage of 147 gross tons.

Owned by Scimitar Trawlers Ltd., *LT100 Ocean Scimitar* was until 1975, the *Boston Scimitar* owned by Aberdeen Near Water Trawlers Ltd. She was built at Richards Ironworks in 1959 and is seen here entering Lowestoft harbour. In 1976 *Ocean Scimitar* was sold to Putford Enterprises Ltd., and renamed *Putford Scimitar*. She was converted for use on safety standby work, and her fishing registry was cancelled shortly afterwards. On the 8th. January 1992, she left Lowestoft after being sold to a new owner on the Isle of Wight.

A rather run down *LT 87 Jadestar Glory,* heads out of Lowestoft on the 16th. August 1973. She was formerly the Talisman Trawlers vessel *Ludham Queen*, built in 1954 by Cochrane & Son at Selby. Owned by H. J. Lamprell in this view, she grounded in 1974, south of Arklow. She was refloated and then towed to Dublin for scrapping. *Ludham Queen* was powered by a 6 cyl. 350hp Crossley engine, and was 102ft long with a gross tonnage of 179 tons.

LT8 St.Thomas makes a fine sight as she heads north off Lowestoft. Built by Richard Dunston at Hessle in 1969, she was powered by a 6 cyl. 1350hp Ruston engine and had a gross tonnage of 241 tons. Her dimensions were 121ft.5ins. x 25ft.1ins. x 11ft. *St.Thomas* was built for Claridge Trawlers and was sold to Anglo-Spanish interests in 1986. She was initially renamed *Pescafish II*. Later this was changed to *Blenheim*.

Built at Appledore in 1967, *LT777 Suffolk Venturer* was built for Small & Co.(Lowestoft) Ltd. One of a class of six, she was 124ft. long and had a gross tonnage of 255 tons. *Suffolk Venturer* left Lowestoft on 7th. April 1987 for a new life with Anglo-Spanish owners, and she was renamed *South Coast*.

Rigged for trawling, *LT387 Young Duke* leaves Lowestoft in 1968. Owned by Small & Co.(Lowestoft) Ltd., she was built in 1953 at Richards Ironworks in Lowestoft. *Young Duke* was sold out of fishing in 1967, but when sold again in 1969, she became *KY377 Spes Aurea*. Later in life she was used on safety standby duties, and also diving support work.

LT402 St.Georges makes an impressive sight, as she leaves Lowestoft on the 26th. June 1975 on another fishing trip. This vessel was destined to achieve world-wide fame when she starred in the James Bond Film " For Your Eyes Only". *St.Georges* was built by Cook, Welton and Gemmill at Beverley in 1946 for J. Marr & Son Ltd. as *H318 Thorina*. Her dimension were 136ft. 8ins. x 25ft. 7ins. x 13ft. 6ins. with a gross tonnage of 343 tons. She was sold by J. Marr & Son in 1948 to Holland, and in 1964 was bought by Claridge Trawlers Ltd. On 7th. June 1984 she left Lowestoft for scrapping at Gravesend.

Built by John Lewis at Aberdeen in 1961, *A530 Hawkflight* was bought by Gorleston owner C.V.Eastick Ltd. in 1973. Her registration was changed to *A215* at the time. She was sold to Huxley Fishing Co. Ltd. in September 1975, and used for fishing as *LT213 Aruba*. Later she was used on safety standby work. On 8th. August 1991, she left Lowestoft after being sold for scrap. The *Hawkflight* was in the ownership of C.V Eastick Ltd. on the 9th. July 1973, when seen here off Lowestoft. As built, she was fitted with a 5cyl. 475hp. National engine and was 98. 6ft. long, with a gross tonnage of 174 tons.

LT501 Yoxford Queen was completed as the stern trawler *SN58 Sailfin* for Pelagic Trawling Co. Ltd., by T. Mitchison Ltd. at Gateshead in 1962. Bought by Talisman Trawlers in 1965, she was converted to a side trawler and given her new name and registration. *Yoxford Queen* was re-engined in 1967 and again in 1969. In 1983 she was sold to Scupham Fishing Co. Ltd. and renamed *Rae-Elizabeth*. On the 29th. June 1976 she is seen here entering Lowestoft. Putford Enterprises Ltd. bought her in 1985, when all useable gear was removed and she was sold for scrap.

(Top left) The Waveney Dock at Lowestoft on a day in 1968 when at least five trawlers landed. The trawlers shown are from the fleets of Talisman Trawlers Ltd., Mitchells Tritonia Ltd., Putford Enterprises Ltd., and Claridge Trawlers Ltd.

(Top right) A view of the Trawl Dock at Lowestoft in the early 1960s. Vessels from the fleets of Dagon Fishing Co., Mostyn & Willy, Eddystone Fishing Co., Aberdeen Near Water Trawlers Ltd., and Clan S.F.(Grimsby) Ltd. can be seen. The vessel in the foreground is one of the twelve trawlers ordered in December 1932 by Mr. W. F. Cockrell, from Richards Ironworks Ltd.

(Bottom right) Looking down from the roof of the former Pryces warehouse at Lowestoft. It is early in the morning of Christmas Eve 1975. On the left, part of the Hamilton Dock can be seen, and on the right, part of the Waveney Dock. Nine trawlers of the Small & Co. (Lowestoft) Ltd. fleet are present. There are three stern trawlers in the Waveney, and all six of the "Appledore" class of side trawlers in the Hamilton.

Scottish fishing vessels visited Lowestoft in late 1973 and again in 1975. They made several landings of herring. These pair trawlers are shown here in the Waveney Dock on the 24th. September 1975.

PD114 Star Crest in seen in the top view, and in the bottom view, *PD90 Accord* is seen being unloaded. Over 300 crans being landed that day. Also taking part in the fishing were *PD111 Trident* and *PD67 Faithful II*.

The Hamilton Dock on the 17th. November 1973, showing the four Scottish vessels. Also seen is the Lowestoft longshore boat *LT202 Cherokee,* and just visible is the stern of *LT446 G. & E.*

The four Scottish vessels in the Waveney Dock on the 15th. November 1973. *PH140 Francis Patrick* is also in the dock. This Great Yarmouth owned vessel was built in France in 1953 and had a gross tonnage of 89 tons.

On completion by Richards Ironworks in 1959, the *Boston Coronet* was registered *GY596*. In 1963 she was re-registered *LT459*. Between 1965 and 1972, *Boston Coronet* was owned by Aberdeen Near Water Trawlers Ltd., an associated company of Boston Deep Sea Fisheries Ltd. (BDSF). From 1959 until 1965, and from 1972 until 1980, she was owned by BDSF. In 1972 *Boston Coronet* was re-engined with an 8cyl. 850hp. Mirrlees, replacing her 5cyl. 550hp Widdop. During 1980 she was sold to Anglo - Spanish interests, and became *Lidia Prima*. This impressive scene was recorded on the 22nd. June 1973.

LT777 Suffolk Venturer was one of the trawlers built at Appledore in the late 1960s for Small & Co.(Lowestoft) Ltd. We see her here turning, and about to enter the Waveney Dock at Lowestoft. During 1982, *Suffolk Venturer* was converted for offshore safety standby duties. On the 7th. April 1987 after being sold, she left Lowestoft with the registration *LT349*; Later she became *AR95 South Coast*.

Putford Enterprises Ltd. is now one of the major leading British operators of offshore support and supply vessels. Many of these offshore vessels are registered at Lowestoft. Based in the town since 1948, the company started operations as fishing vessel owners. At one time they had a large fleet of drifter/trawlers and trawlers operating from the port. Two of their later trawlers are featured here.

Built for the company in 1960 by Richards Ironworks Ltd. as yard no. 452, *LT240 Woodleigh* was powered by a 6 cyl. 550hp. Crossley engine. Her gross tonnage was 199 tons and she had dimensions of 106ft. 6ins. x 23ft. 4ins. x 9ft. 7ins. In 1975 *Woodleigh* was converted for use as a safety standby vessel for the offshore oil and gas industry. She was sold during 1993, reportedly for future use as a restaurant. In 1997 the vessel was seized during an anti-drug operation, and taken to Falmouth. On the 9th. October 1973, we see her entering her home port.

Previously owned by Small & Co.(Lowestoft) Ltd. as *LT422 Suffolk Craftsman*, the *Winkleigh* was also a product of Richards Ironworks Ltd. Completed in 1962 as yard no. 462, she was powered by a 6 cyl. 550hp Ruston engine. Her dimensions were 106ft. 6ins. x 23ft. 4ins. x 12ft. 3ins. She had a gross tonnage of 202 tons. Bought by Putford in 1974, she was converted for safety standby work in 1981. *Winkleigh* sailed on the 6th. August 1993 from Lowestoft, reportedly bound for Madagascar after being sold. She is seen here on the 21st. June 1975 entering Lowestoft harbour.

A number of distant water trawlers moved to Lowestoft in the late 1970s and early 1980s. Many came from ports such as Aberdeen, Fleetwood and Grimsby.

Known to many as *LO72 Captain Riou,* this trawler became *FD173 Boston Defender* in 1972. Seen here leaving Lowestoft on the 28th. August 1974, she was owned by the Iago Steam Trawler Co. Ltd. During 1976, *Boston Defender* was converted for the offshore safety standby role and her fishing registration cancelled. She was sold to George Craig & Son Ltd., Aberdeen during 1978. *Captain Riou* was built in Aberdeen by John Lewis in 1957 as yard number 268. When completed, she was 139ft. 6ins. long and had a 5cyl. 800hp British Polar engine.

An unusual sight at Great Yarmouth on the 10th. August 1978, as the trawler *GY622 Gillingham,* moves down the river. Owned by Huxley Fishing Co. Ltd., Lowestoft, this 137ft. long vessel had just spent some time in the dry dock. Her registration was changed to *LT305* in late 1978. Used for a number of years for fishing from Lowestoft, she was eventually converted for offshore safety standby duties. Built in 1960, *Gillingham* was previously part of the Consolidated Fisheries fleet. She left Lowestoft for a shipbreakers yard on the Medway on the 8th. July 1987.

Although never to fish from Lowestoft, *GY694 Northern Reward* was technically still a fishing vessel when she arrived at the port. Bought from British United Trawlers Ltd. by Colne Shipping Co. Ltd., she became the safety standby/ hospital vessel *St. Elizabeth* in 1982. Colne Shipping sold her in July 1992. Her new owner continued to employ her on offshore safety work. In the late 1990s she was a regular visitor to Great Yarmouth, along with other former distant water trawlers. *Northern Reward* was built by Cook, Welton & Gemmell at Beverley in 1962. When built she was 164ft. long with a gross tonnage of 576. She is seen here crossing the Waveney Dock with the Colne Shipping Co. tug *Eta* in charge, on the 24th. January 1981.

Built in 1962 at Goole, *GY702 Huddersfield Town* was another vessel to find a new home at Lowestoft. She was for many years part of the Consolidated Fisheries fleet at Grimsby. In July 1978, *Huddersfield Town* was bought by Huxley Fishing Co. Ltd. and became *LT259*. Used initially for fishing from Lowestoft, she was later converted for use on offshore standby work. In this splendid view, we see *Huddersfield Town* leaving Lowestoft on the 27th. September 1979 on a fishing trip. On the 10th. January 1992, she left Lowestoft after being sold for scrap.

Between 1972 and 1974, Small & Co.(Lowestoft) Ltd., took delivery of four new stern trawlers. These vessels were much admired, but had a very short working life as trawlers. They were built by Cubow Ltd., Woolwich, and were 130ft. long and had two 1000hp. Blackstone engines. One of the four was *LT175 Suffolk Harvester*, seen here off Lowestoft on the 9th. April 1973. In 1978, this vessel and another of the class were chartered to the Ministry of Defence for use as mine countermeasure trials ships. On return to Lowestoft at the end of 1983, both vessels were converted for offshore support work and joined the other members of the class. In 1990 all four vessels were sold. The names were changed by replacing the "Suffolk" prefix of the four, with "Britannia". The *Suffolk Harvester* becoming *Britannia Harvester*. These four former stern trawlers can be seen occasionally in Lowestoft and Great Yarmouth in their new guise as offshore safety standby vessels.

The Grimsby trawler *GY644 Judaean* is seen here at Lowestoft on the 25th. November 1976, the day of her arrival at the port. She was bought by Dagon Fishing Co. Ltd., an associate of the Colne group, from Sir Thomas Robinson & Sons(Grimsby) Ltd. Delivered to Lowestoft still essentially as a fishing vessel, her appearance and name were soon to change. *Judaean* was converted for offshore safety standby work and became *Abaco*. In July 1984 she caught fire and was seriously damaged.

Looking very smart, the former Fleetwood trawler *Boston Phantom* waits in the bridge channel at Lowestoft on the 18th March 1982. Converted for use on offshore safety standby work in 1979, she was bought by Claridge Trawlers Ltd. in 1984. During April 1985, the *Boston Phantom* became *Colne Phantom,* a unit of the large fleet of standby vessels operated by the Colne group. She carried out offshore safety standby duties until 1992. *Colne Phantom* was sold to Caravelle Maritime at Rainham, and sold on to South Africa, reportedly to revert to fishing. *Boston Phantom* was built in 1965 at Beverley and was 140ft. in length with a gross tonnage of 431. She first arrived at Lowestoft on the 7th. December 1979.

After being transferred from Grimsby, the stern trawler *GY321 Boston Halifax* fished for part of 1979 out of Lowestoft. She was then laid up. On 14th. September 1979 she sailed to Norway for conversion to pair trawling. The intention was for her to work with *Boston Stirling,* out of Fleetwood. During 1985 *Boston Halifax* was re-registered *LO339*. In 1986 she was sold to Denmark and became *E130 Drot*. During 1991, she was sold to George Craig & Sons Ltd. and became *Grampian Dee. Boston Halifax* was built by the Goole Shipbuilding & Repair Co. Ltd. in 1975, her dimensions were 123ft. 11ins. x 31ft. 1ins. x 12ft. 5ins. She had a gross tonnage of 387 tons, and her main power plant was two 8cyl. 1440hp. Mirrlees Blackstone engines.

GY465 Parkroyd was completed in 1960 by Vosper's at Portsmouth. She was bought in 1976 by the Colne Shipping Co. Ltd., when she became *LT251 St.Croix*. Her previous owners had been Near Water Trawlers Ltd., Fleetwood, and North Star Steam Fishing Co. Ltd., Aberdeen. Whilst at Aberdeen she was registered *A161*. As built, the *Parkroyd* was 113ft. long and of 310 gross tons. In January 1986, the *St.Croix* left Lowestoft for a scrapyard at Gravesend. We see her here entering the harbour at Lowestoft, under tow, on the 19th. February 1977.

The Waveney Dock at Lowestoft in the early 1970s.

LT500 Barnby Queen was one of the two stern trawler built at Goole for Talisman Trawlers Ltd. She was completed in 1976. *Barnby Queen* was 124ft. in length and had a gross tonnage of 349 tons. She was sold to Denmark, and left Lowestoft on the 7th. July 1984. There she became *HG142 Nordorn*. She is seen here on the 5th. July 1984, approaching Lowestoft Bridge. Five months later in December 1984, the *Nordorn* sank with serious loss of life.

The fishing fleets of Lowestoft and Great Yarmouth have always consisted of a wide variety of craft and vessels. Most books on the fishing industry tend to overlook the longshore boats and inshore fishing vessels. Here are three representatives of that important side of the industry.

(Top Left) An example of the many beach launched boats which have Gt.Yarmouth and Lowestoft registrations. With a launching tractor behind her, is *YH78 Rosebay*. This scene is at Caister on the 23rd. July 1991, and also shows a lifeboat of the very famous independent Caister Volunteer Rescue Service. The lifeboat is the *Shirley Jean Adye*. She had just been replaced by a new lifeboat, and had been in service at Caister from 1973 until 1991. The Caister station was closed by the R.N.L.I., but reopened by local people.

(Top right) *YH223 Alida* heads back to port on the 2nd. January 1978. She was built in 1958 at Whitby.

(Bottom Right) The Danish built *YH541 Roannah,* seen off Great Yarmouth. She was built in 1961.

(Left) An impression of *YH77 Ocean Dawn* as delivered by Richards Ironworks to Bloomfields Ltd. She is off Lowestoft undergoing trials.

(Below) The date is the 23rd. April 1984, and work is well under way at Oulton Broad in converting *Ocean Dawn(Rewga)*, into her new role as a SSV. Close examination will reveal part of her last fishing registration, *KY371*.

YH77 Ocean Dawn

YH77 Ocean Dawn was one of the six drifter/trawlers built between 1952 and 1957 for Bloomfields Ltd. of Great Yarmouth, by Richards Ironworks Ltd., at Lowestoft. She was sold in 1963 to the East Anglian Ice & Cold Storage Co. Ltd. and on the 21st. February 1963 was re-registered *LT466*. During 1968, she was transferred to Small & Co. (Lowestoft) Ltd. On the 2nd. September 1969 she sailed from Lowestoft, after being sold, bound for Scotland. There she was re-registered *KY371*. *Ocean Dawn* was under Scottish ownership until 1984, when her owner Mr. John Muir sold her back to her birth place. At Lowestoft under the ownership of the Colne Shipping Co. Ltd. she was converted, like so many other drifters and trawlers, for use as a safety standby vessel (SSV). Her name was changed from *Ocean Dawn* to *Rewga*. She was sold in 1987 and left Lowestoft on the 24th. February, bound for Sweden. Her new owner changed her name back to *Ocean Dawn*. The *Ocean Dawn* was completed in February 1956, and underwent trials in the first week of March. On the 5th. March, she could be found off Great Yarmouth undergoing compass testing and adjusting. This 133 gross tons vessel, had dimensions of 91ft. 3ins. x 20ft. 7ins. x 9ft. 7ins., and her main engine was a 360hp 6 cyl. Ruston. The first skipper of the *Ocean Dawn* was Mr. E. Harris.

LT173 Boston Hornet was the last drifter/trawler built for Boston Deep Sea Fisheries Ltd. She was completed in May 1960 at the Brightlingsea yard of James and Stone (Brightlingsea) Ltd. With dimensions of 91ft. 7ins. x 22ft. 4ins. x 10ft. 6ins., she had a gross tonnage of 131 tons. For many years used as a SSV, the *Boston Hornet* received national publicity when she saved many lives from the sinking SSV *Hector Gannet*. The *Hector Gannet* sank after colliding with the rig *Hewett A*, during an emergency evacuation of the rig. The *Boston Hornet* was sold, and left Lowestoft on the 30th. July 1986 bound for Spain. She was later sold on, and by 1992, had been converted into a yacht. Mid morning on the 21st. September 1974, the *Boston Hornet* is seen here leaving her home port for another spell of duty as a SSV.

Another of the many Lowestoft fishing vessels to be sold to Anglo - Spanish interests was *LT7 St. John*. Built by Richard Dunston Ltd. at Hessle in 1969, her dimensions were 121ft. 6ins. x 25ft. 10ins. x 11ft. She had a gross tonnage of 241 tons, and was powered by a 1350hp. 6cyl. Ruston. In October 1986 she was sold and became *Pescafish 1*, and in 1991 became *Badminton*. The *St.John* is seen here on the 9th. August 1975, heading out of Lowestoft.

A scene suitable for the front of a Christmas card, showing *LT502 Farnham Queen* at Lowestoft on the 18th. February 1979. As the Anglo-Spanish vessel *Sea Dog,* she sank 200 miles off the coast of Ireland on the 16th. December 1998.

Powered by a 1095hp. 6cyl. Ruston engine, *LT139 Boston Sea Fury* leaves Lowestoft on the 9th. August 1975, for the fishing grounds. In 1984, she was used on SSV duties. Later she was sold to Britannia Marine Ltd., and in 1988, became the SSV *Britannia Fury.* On the 31st. December 1993, she left Lowestoft, reportedly bound for Spain, after being sold on. Her name was shortened to *Fury.* Built by H. McLean & Sons Ltd. at Renfrew in 1973, she had dimensions of 109ft. 1ins. x 26ft. 11ins x 10ft. 6ins. *Boston Sea Fury* had a gross tonnage of 312 tons.

One of the many trawlers which were bought by various Lowestoft owners to operate from the port in the 1970s, was *GN72 Granton Merlin*. She was built in 1960 by Hall Russell Ltd. at Aberdeen. After having a number of different owners including British United Trawlers, she was bought by Putford Enterprises Ltd. At Lowestoft, she was converted for use as a SSV, and became the *Umberleigh*. As built, *Granton Merlin* was powered by a 705hp. 5cyl. Ruston engine., and had dimensions of 108ft. x 24ft 1ins. x 12ft. In this view, she is passing the Co-operative Wholesale Society canning factory at Lowestoft. It is the 6th. February 1978. At the time of writing the *Umberleigh* is in the process of being converted into a luxury yacht at the former Oulton Broad shipyard of Brooke Marine Ltd.

One of the two stern trawlers built for Claridge Trawlers Ltd. at Gt.Yarmouth in 1975, *LT144 St. Phillip* was sold in September 1986, and became *Kerry Kathleen*. She returned to Lowestoft ownership in 1989, and regained her previous name. *St. Phillip* was converted to a SSV shortly afterwards. Following a number of further changes in ownership, this vessel became *Viking Vulcan* in 1998. As a stern trawler, we see her entering her home port on the 28th. June 1976.

As mentioned on page 157, Putford Enterprises have maintained a major presence at the port of Lowestoft since 1948. Initially fishing vessel owners, the company now operates safety standby, and supply vessels, in support of the offshore oil and gas industry. Units of their large fleet can regularly be seen in port at Lowestoft and also Great Yarmouth. Two of their previous vessels are featured here.

A377 Dreadnought was built in 1960 by Herd & Mackenzie Ltd. at Buckie. Her main engine was a 454 hp 6cyl. Ruston, and her dimensions were 93ft. 9ins. x 22ft. 4ins. x 10ft. She was bought by the Company in 1970 from the White Fish Authority. Initially used for fishing, *Dreadnought* was also used as a SSV. She is seen here as a trawler in the Trawl Dock at Lowestoft. In 1975 she was officially converted for use as a SSV and her fishing registration cancelled. Her name was changed to *Putford Harrier* in 1984, thus bringing her into line with the naming policy used today for Putford vessels. In addition to use as a SSV, this vessel was also used on aircraft crash recovery work. *Putford Harrier* was sold in May 1992.

Leaving Lowestoft in the late afternoon is *FD42 Westleigh*. Still technically a trawler, she was employed at the time on support work. This large vessel was the former Fleetwood trawler *Boston Seaform*. She was built in 1956 for Boston Deep Sea Fisheries Ltd. by Richard Dunston Ltd. at Hessle. Her name was changed to *Westleigh* in 1974 when bought by the Company. She had been fishing from Lowestoft since 1970. During 1976, *Westleigh* was sold and became the SSV *Arkinholme*. Sold again in 1978, this time to George Craig & Son Ltd., she had a further name change, and became *Grampian Castle* in 1979.

Sold in 1986 to Clipper Promotions, with the intention of being converted into a sailing ship, the *Idena* started life in 1953 at the shipyard of Cook, Welton & Gemmell at Beverley. She was built for J. Marr Ltd., Fleetwood and when completed had the registration *FD136*. Her dimensions were 135ft. 8ins. x 26ft. 8ins. x 13ft., and her main engine was a 773hp. 7cyl. Mirrlees. During 1967, she was transferred to the Ranger Fishing Co. Ltd. with her registration changing to *A793*. Bought by Putford Enterprises Ltd. in 1971, *Idena* became *Falkirk* in 1974. Her fishing registration was closed and she was fully converted for SSV duties. *Falkirk* passed to Safety Ships Ltd. in 1975. During 1980 she had a further name change and became *Grampian Falcon,* a unit of the George Craig & Sons Ltd. fleet. We see her here leaving Lowestoft for a spell of duty as a SSV, prior to full conversion, and still essentially a trawler.

The Waveney Dock at Lowestoft in the late 1960s, with icing in progress.

Demolition in progress of the large marketing and processing hall on the west side of Waveney Dock in 1984. This magnificent building, built by the Great Eastern Railway(GER), resembled a great railway station. The many merchants offices at the back of the hall closely resembled GER signal boxes. The bottom right view, on page 153, illustrates these offices very well. A new hall was built on the same site. Two of the large Lowestoft fleet of SSVs can be seen, they are the former trawlers *SSAFA* and *Notts Forest*.

Sold for Scrap

Steam and diesel powered vessels considered to be at the end of their useful lives. These scenes show Scottish and English vessels after disposal by their owners. They show typical final resting places, and also a typical last voyage made by many other fishing vessels.

The steam drifter *LT730 Implacable* at a shipbreakers yard in Oulton Broad. Either side of her are the remains of two similar vessels. The *Implacable* was in very good condition when sold for scrap, and was previously owned by Metcalf Motor Coasters Ltd. She was built at the John Chambers shipyard, almost opposite where she is seen in this view, in 1916. The *Implacable* moved to this, her last berth, on the 7th. April 1955.

The scene at Bo'ness on the 3rd. June 1980, with the trawler *A491 Glengairn* awaiting the cutters torch. She was built at Aberdeen in 1960, and had a gross tonnage of 228 tons.

Leaving Lowestoft for the last time in October 1954, for a breakers yard at Antwerp is the steam drifter/trawler *LT273 Lord Duncan*. She was built in 1920 as the standard drifter *HMD Melody* and became *BCK122 Rose Duncan,* and later *SH105.* On the 7th. July 1954, *Lord Duncan* sank in the Inner Harbour at Lowestoft. Her last local owner was Lowestoft Herring Drifters Ltd. *Lord Duncan* was towed to the breakers by the steam trawler *LT572 Ouse,* also making her final voyage.

Taken at the same location and date as the view featuring *Glangairn* on the opposite page, we also see *A472 Milwood* similarly waiting to be cut up. The *Milwood* was built in 1960 at Gateshead. In both views of this shipbreakers yard, a number of other side fishing trawlers can be seen awaiting a similar fate.

Two aspects of the changing scene at Lowestoft in the 1970s and 1980s, was the increasing number of side fishing trawlers being used as safety standby vessels (SSVs), and the re-introduction of beam trawling from the port.

The scene from the yard of Richards(Shipbuilders) Ltd., with two SSVs present. Both vessels are former side trawlers owned by Boston Deep Sea Fisheries Ltd. Opposite can be seen *LT145 Wilton Queen,* owned by Talisman Trawlers Ltd. Built by Richards in 1960 as yard no. 455, she was sold in November 1981 to Anglo-Spanish interests. As the *Contessa Viv* she sank in August 1986.

An early arrival in the beamer revolution at Lowestoft was *GO7 Jacob.* This 1974 built vessel was bought by the Colne Shipping Co. Ltd. in March 1984, and became *LT59 St.Georges.* On the 20th. June 1984, we see her leaving her new home port on a fishing trip. During January 1995, *St.Georges* was sold and re-registered *PZ1053*. Prior to delivery to her new owners, she visited Holland for engine repairs. She was towed there by a Dutch tug, leaving Lowestoft on the 15th. January 1995.

Sold to Anglo-Spanish owners in February 1987, the 255 gross tons trawler *LT789 Suffolk Endeavour* is seen off Lowestoft in early 1970s heading for the fishing grounds. Built in 1968 at Appledore she was owned by Small & Co.(Lowestoft) Ltd. In 1977 she was re-registered *LT264,* and from 1976 until her sale in 1987, was used on offshore safety standby work. She left Lowestoft as *LT374.*

The 190 gross tons stern trawler *LT293 Boston Sea Stallion* was built in 1978 by Richards(Shipbuilders) Ltd. at Great Yarmouth. Powered by a 700hp 8cyl Mirrlees engine, her dimensions were 86ft. 2ins. x 25ft. 10ins. x 10ft. 7ins. Her owners, Boston Deep Sea Fisheries Ltd., had her converted in 1984 for use as a SSV and her fishing registry was cancelled. In 1988, with transfer of ownership, she became *Britannia Stallion.* She was sold on in April 1992, and became *Kos Venturer.* In this truly superb view she is undergoing trials off Great Yarmouth shortly after completion.

This very attractive scene shows *LT418 Boston Sea Harrier* at Ullapool in 1980. Owned by Boston Deep Sea Fisheries Ltd., she was built by Richards (Shipbuilders) Ltd., at their Great Yarmouth yard in 1979. This 313 gross tons stern trawler, had dimensions of 109ft. x 30ft. 3ins. x 13ft. 3ins. During 1984, *Boston Sea Harrier* was re-registered *LO338*. She was later sold to Norway and renamed *Sylvester*. During 1992, *Sylvester* was sold and became *Ryving*. In 1994 she was sold on, and once more renamed.

Any book which specialises in the fishing industry of the British Isles would not be complete, without mention of the renowned *LT30 Ripley Queen*. She was the last large operational side fishing trawler built on traditional lines, to work on the east coast. The full story, together with colour photographs, of her last few weeks in 1994 are to be found in the book "Down The Harbour 1955-1995". On the 12th. June 1973, she is leaving Lowestoft, her home port for her complete life, for the fishing grounds. She was owned by Talisman Trawlers (North Sea) Ltd.

One of the many trawlers to become surplus to requirements at the Humber ports in the 1970s and 1980s was the large freezer trawler *H385 Sir Fred Parkes*. The vessel was acquired in 1982 by the well known offshore support specialists Putford Enterprises Ltd. The *Sir Fred Parkes* was bought with the intention of converting her into a diving support vessel. She was however initially employed as a guard ship under contract to Trinity House. Her job, along with other Putford vessels, was to warn off shipping whilst the laying of electricity interconnector cables was carried out between the U.K. and France. With completion of this contract she returned to her home port of Lowestoft. Later she was to return to fishing with the local name *Waveney Warrior,* and her registration was changed to *H39.* After a period of fishing U.K. waters, *Waveney Warrior* was sold to Spanish owners, and fished the waters around the Falklands. She was sold in 1992 to a company in Argentina and renamed *Corcubion. Sir Fred Parkes* was built in 1965 by Hall, Russell & Co. at Aberdeen. She is seen here off Lowestoft with her Guard 1 markings ready for work in the Channel, retaining her Hull fishing registration..

The 24 gross tons *LT299 Waterloo Warrior* seen at Pittenweem on the 8th. March 1980. This vessel was built in 1971. Until 1978 she was *Christophe Stephane*.

The conversion of fishing vessels into support vessels for the offshore oil and gas industry has been mentioned a number of times in this book. The great majority of these vessels arrived at Lowestoft as trawlers. The work of converting them for safety standby duties was carried out in the port, and in some cases at Great Yarmouth.

One such vessel was the Hull trawler *H261 Lord St.Vincent*. She was bought in 1980 by the Colne Shipping Co. Ltd., and converted into a Safety Standby Vessel (SSV) with enhanced medical facilities. Her name was changed to *St. Anne*. At Lowestoft, in the Inner Harbour on the 25th. July 1991, the *St. Anne* waits for her next spell of duty, providing safety services in the gas and oil fields. *St. Anne* was sold in December 1992. The *Lord St.Vincent* was built in 1962 at Beverley. Her previous owner was Hellyer Bros. Ltd. She was of 594 gross tons and 166ft. in length.

The newly completed beam trawler *P225 Cromer* undergoing trials off Portsmouth during July 1988. This 377 gross tons vessel was bought by Colne Shipping Co. Ltd in late 1993. She arrived at her new home port of Lowestoft on the 31st. December 1993. Before being renamed, she completed a number of fishing trips from Lowestoft, the last being on the 22nd. March. *Cromer* was renamed *St.Lucia* in April 1994. The *St. Lucia* remains an important unit of the Lowestoft trawling fleet today, still retaining her Portsmouth registration.

East Anglian based inshore fishing vessels

LT122 De Vrouwe Blanche was bought by Gallidoro Trawlers Ltd. in 1987 as *GO22 Cornelis*. This 203 gross tons beam trawler was 108ft. in length. *De Vrouwe Blanche* was sold in 1993 and became *OB122 Ocean Spirit*. She was sold on during 1993. In the late 1990s she was sold to W. Stevenson & Sons and became the *Cornishman*.

Seen here at Scrabster on the 26th. July 1995, *LT197 Cleopatra* was formerly *Z97 Jakoba*. She first arrived at Lowestoft on the 17th. January 1994. This 273 gross tons vessel is 114ft. in length. and was built in Belgium during 1986.

AT THE END OF THE TWENTIETH CENTURY

9th. December 1998. In the morning, dense fog covered parts of the North Sea. Visiting beamer *N181 Margaret-C* slowly approaches the entrance to Lowestoft.

25th. April 1998. Another of the many visiting fishing vessels to be seen at Lowestoft, the beam trawler *GY527 Fokke* arrives in the early evening.

16th. October 1998. Entering Great Yarmouth harbour in the early morning is *YH1 Our Seafarer*, one of the port's very active inshore fleet.

29th. December 1998. A unit of the present Lowestoft fleet, *LT87 St. Thomas* leaves port after the Christmas break. She is owned by the Colne Shipping Co. Ltd.

26th. December 1995. Colne beam trawlers in port for Christmas at Lowestoft.

27th. December 1993. Talisman Trawlers Ltd. beamers at Lowestoft North Quay.

7th. August 1997. *LT88 St. John* approaching Lowestoft Bridge from Riverside.

LT964 Dryfeholm Queen leaving Lowestoft harbour. She was sold for scrap in 1997

Lowestoft 1999. Vessels landing in the Waveney Dock.

BIBLIOGRAPHY

Down The Harbour 1955-1995	White	White
40 years of fishing vessels, owners,		
the harbour and shipyards at Lowestoft		
Deep Sea Trawling and Wing Trawling	Gourock	Gourock
Fishing Vessels of Britain	Various Editions	EMAP
Fishing Vessels of Lowestoft	Duncan	Woodside
From Tree to Sea	Ted Frost	Dalton
Herring Heydays	K. W. Kent	S.B.Publications
Lloyds Register of Shipping	Various Editions	Lloyds
Maritime Directories	Various Editions	H.M.S.O.
Maritime Great Yarmouth	P.Allard/Parry Watson	S.B.Publications
Merchantile Navy List	Various Editions	H.M.S.O.
Olsens Almanack	Various Editions	Dennis
PLRS Newsletters	Various Editions	PLRS
The Driftermen	Butcher	Topsail
The First Hundred Years	Goodey	Boydell
The Fishing Port of Lowestoft-1st Edition	E.H.C.A.	Burrow
The Fishing Port of Lowestoft-2nd Edition	E.H.C.A.	Burrow
The Lowestoft Fleet List 1974	King	PLRS
The Trawlermen	Butcher	Topsail
Trawling	Hodson	Hodson